ADVANCING ENVIRONMENTAL EDUCATION PRACTICE

A volume in the series
Cornell Series in Environmental Education
Edited by Marianne E. Krasny

For a list of books in the series, visit our website at
cornellpress.cornell.edu

ADVANCING ENVIRONMENTAL EDUCATION PRACTICE

Marianne E. Krasny

COMSTOCK PUBLISHING ASSOCIATES
AN IMPRINT OF CORNELL UNIVERSITY PRESS ITHACA AND LONDON

First published 2020 by Cornell University Press

Library of Congress Cataloging-in-Publication Data

Names: Krasny, Marianne E., author.
Title: Advancing environmental education practice / Marianne Krasny.
Description: Ithaca : Comstock Publishing Associates, an imprint of Cornell University Press, 2020. | Series: Cornell studies in environmental education | Includes bibliographical references and index.
Identifiers: LCCN 2019020285 (print) | LCCN 2019021611 (ebook) | ISBN 9781501747076 (pbk.)
Subjects: LCSH: Environmental education. | Environmental education—Social aspects.
Classification: LCC GE70 .K73 2020 (print) | LCC GE70 (ebook) | DDC 363.70071—dc23
LC record available at https://lccn.loc.gov/2019020285
LC ebook record available at https://lccn.loc.gov/2019021611

ISBN 978-1-5017-4708-3 (PDF ebook)
ISBN 978-1-5017-4709-0 (epub/Mobi ebook)

To the members of the Civic Ecology Lab—
Our work together has contributed happiness and meaning to my life

Contents

Preface

For years, environmental education has faced an existential crisis. Is its goal to build the capacity of participants to make their own decisions about whether or not to take action? Or do environmental educators aim to improve the environment—using education as a tool to address the environmental crisis? Recently, as research documenting the link between nature and human well-being has captured the public's attention, environmental educators have turned to health and youth development outcomes for program participants. Other environmental educators view their work as science literacy—as a means to make science come alive for students and to teach about ecological systems.

As the author of this book, I am not shy about my belief that environmental education should be directed at addressing environmental problems. But I have tried to write for those who prioritize building participants' capacity to make informed decisions, helping humans thrive, and teaching science. I believe that we benefit from a "big tent" approach that encompasses a diversity of ideas and strategies. I also believe that different environmental education programs can realize multiple goals simultaneously—and that the pathways to realize environmental quality, community well-being, and human health are intertwined.

But—and this is the thesis underlying this book—environmental educators, regardless of our ultimate goals, are more effective when we articulate sound *theories of change*. A theory of change is our beliefs about how program activities lead to program outcomes. We all have big goals like environmental quality, sustainability, resilience, or ensuring that every child thrives. To get to these big, or ultimate, outcomes of environmental education, people need to engage in environmental behaviors, like reducing meat or energy consumption, and in collective action, like restoring wetlands or advocating for a carbon tax.

Environmental educators impart knowledge, influence attitudes and norms, nurture environmental identity and political efficacy, and build trust in order to change participants' behaviors. A theory of change drawing from research on how trust enables people to act collectively might be "When participants do a challenging outdoor activity like climbing a mountain or running a race together, they learn to trust each other. Having built trust, they are more likely to work together to help steward a local green space." Other times environmental education programs start by engaging participants in the actual desired behaviors—a nature center serves only vegan meals, or a business incentivizes workers to volunteer for a litter cleanup. The environmental educator might reason, "Through engaging in a litter cleanup, participants develop norms that will lead to similar future behaviors. By the time they pick up the twentieth plastic straw, maybe they start to realize connections with their own behavior and responsibility."

Whether their pathway to changing behavior is through first building trust that leads to collective action, through engaging participants in a behavior where they develop norms that lead to future behaviors, or any number of other pathways, environmental educators can articulate their theory of change. Generally, a theory of change is expressed as a diagram showing the relationships among program activities; intermediate outcomes like trust, norms, knowledge, and efficacy; environmental behavior or collective action outcomes; and even ultimate outcomes like environmental quality, health, and resilience. A short narrative explaining the diagram is also part of a theory of change.

All environmental educators—including myself—can do more to critically reflect on how our theories of change determine what we do. I think that the lack of attention to our theories of change, and the tendency to do what seems natural in education—that is, to teach or convey knowledge—are even more important than the differences we debate about the fundamental purpose of environmental education. We can incorporate multiple approaches, but to reach our goals, we need to pay attention to the research and our experience, and to articulate, test, and adapt our theories of change. We also need to realize that environmental education is one tool in a toolbox—or perhaps more accurately one node in a network—of interwoven efforts to steward, restore, and even transform our environment in ways that benefit humans and other life. In addition to environmental education organizations, nodes in the network include NGOs focused on community and youth development, businesses seeking to address sustainability, and universities wanting to engage in participatory research to solve real problems. Environmental education alone cannot address the climate crisis, plastics proliferation, or human health. But it can play an important role working alongside—and by linking with—other social and environmental endeavors.

Acknowledgments

I thank Anne Umali, Elizabeth Danter, and Alex Kudryavtsev, who contributed to earlier writing about outcomes of environmental education. I also thank Alex Kudryavtsev, Yue Li, and Anne Armstrong for many years of conversation and dedicated work to advance environmental education. This publication was funded in part by the US Environmental Protection Agency (EPA, Assistant Agreement No. NT-83497401) and the US Department of Agriculture (USDA) National Institute for Food and Agriculture funds awarded to Cornell University (Award No. 2016-17-215). Neither the EPA nor the USDA has reviewed this publication. The views expressed are solely those of the author.

ADVANCING ENVIRONMENTAL EDUCATION PRACTICE

Introduction

My ultimate goal in writing this book is to better position environmental educators to contribute to environmental quality, sustainability, and resilience. To accomplish this goal, I have summarized research-based information on the myriad pathways by which environmental education can contribute to the health of the environment, the community, and individuals. Like other researchers, I challenge the knowledge-attitudes-behavior pathway—the assumption that environmental knowledge and attitudes lead to environmental behaviors. Instead I review research that shows that certain types of knowledge are more likely than others to influence behaviors, and that sometimes it is better to work with existing attitudes than to try to change them. I then expand our purview of potential intermediate outcomes of environmental education beyond knowledge and attitudes to include nature connectedness, sense of place, efficacy, identity, norms, social capital, youth assets, and well-being. All these intermediate outcomes can be nurtured through environmental education and can lead to future environmental behaviors and collective action.

Environmental education encompasses any learning activities that help ecosystems and societies thrive. It includes learning opportunities embedded in hands-on stewardship, citizen science, environmental activism, and unstructured time spent in nature. And it is part of a larger effort by policy makers, researchers, the private sector, and civil society to respond to pressing environmental challenges. The goal of environmental education is nurturing individual behaviors and collective actions that lead to healthy and resilient environments and communities.

Whether you are a practicing or prospective environmental educator, I hope you will benefit from the synthesis of environmental psychology and related research found in the chapters of this book. Whether you work at a nature or community center, national park or urban pocket park, botanic garden or community garden, in the school classroom, or in a museum, aquarium, or zoo, the information in this book should help you home in on ways you can most effectively engage and influence your participants. In addition to the environmental educator, this book is written for the college student volunteering in an environmental club or considering pursuing an environmental education career. Whether you want to directly improve the environment, to enhance systems knowledge and critical thinking, to create environmental norms and social capital, or to foster youth development, you should be able to find information in this book to help you achieve your goal.

In the beginning chapters of this book (part 1), I focus on theory of change and evaluation strategies. I turn next to exploring environmental quality outcomes of environmental education, followed by separate chapters on individual behaviors and collective action (part 2). Whereas individual behaviors and collective actions are often hard to separate in environmental education programs, the research on factors leading individuals to change a behavior differs in important ways from findings on what influences a group of people to take action together. For example, self-efficacy plays a prominent role in what we do as individuals, whereas collective and political efficacy and social capital play a role in what we do as a collective. In places where there is overlap between behaviors and action, I use the two terms interchangeably, whereas in sections where I discuss factors specific to individuals or groups, I distinguish between individual behaviors and collective action.

To reach its ultimate goal of improving the environment, environmental education fosters action-related knowledge and systems thinking, takes attitudes into account in program planning, and provides opportunities to connect with nature and with place. It helps people develop feelings of efficacy and forge environmental identities. Environmental education programs can set the standard for environmentally friendly norms and create social capital among participants and community members. And environmental educators engage youth in activities that foster positive development, health, and well-being, including a sense of hope. All these intermediate outcomes (part 3) can be viewed as cognitive and affective pathways to environmental behaviors and action. Sometimes, environmental education programs consider youth development, well-being, or another intermediate outcome as their most important goal, focusing less on environmental behaviors per se. Finally, environmental education can start by engaging participants in stewardship or other action, which in turn fosters the cognitive and affective intermediate outcomes that then lead to additional environmental behaviors. Although many focus on changing the way people think and feel in order the change their behavior, it is important to keep in mind that performing a behavior can also change the way we think and feel.

In short, this book helps educators to plan, assess, adapt, and transform their programs based on research findings. It does not offer specific instructions for lesson plans or activities, which can be found in numerous publications produced by nonprofit organizations such as Earth Force, the Paleontological Research Institute, or the Nature Conservancy, as well as by the US Forest Service and Association of Fish & Wildlife Agencies. Nor does it describe the wealth of environmental education practices. For a compendium of contemporary environmental education practices, Cornell University Press's *Urban Environmental Education Review* may be useful (Russ and Krasny 2017).

Below is a brief overview of the individual chapters, followed by discussions of the controversy surrounding environmental quality as the ultimate goal of environmental education and of the largely discounted knowledge-attitude-behavior theory of change. I close this introductory chapter with reflections on environmental education as one node in the network of endeavors addressing environmental quality.

Chapter Summaries

Read a quick summary of each chapter below.

Theory of Change (Chapter 1)

A theory of change is a diagram and narrative that shows how your program activities lead to your intermediate, behavior/action, and ultimate outcomes. As the Cheshire Cat observed in *Alice in Wonderland*, if you don't know where you are going, any road will take you there. In other words, without a theory of change, an environmental education program "is vulnerable to wandering aimlessly" (Reisman and Gienapp 2004, 1).

Evaluation (Chapter 2)

Evaluation presents an opportunity to revisit our theories of change, initially to specify outcomes to evaluate and later to adjust our activities, outcomes, and theory of change based on informal observations and formal research. Whereas environmental educators sometimes conceive of evaluation as an unwelcome obligation or an opportunity to prove their success, a "learning through evaluation" culture spurs program change when needed. In addition to pre-/post- surveys, evaluation can encompass learning activities embedded in programs as well as "Most Significant Change" and appreciative evaluation strategies that focus on how a program achieves positive outcomes.

Environment, Sustainability, and Climate Change (Chapter 3)

For many environmental educators, the ultimate goal is to improve environmental quality. Environmental education can improve environmental quality directly, for example by restoring pollinator habitat or decreasing greenhouse gas emissions. Environmental education also can have an indirect impact through working to change resource management practices and policies. Other approaches to address environmental issues, such as sustainable development and resilience, integrate social and economic alongside environmental concerns.

Environmental Behaviors (Chapter 4)

Lifestyle behaviors, like taking shorter showers, recycling, or turning down the heat, often come to mind when we talk about environmental behaviors. But environmental behaviors are much broader than what we do in our home or workplace. They include hands-on stewardship, teaching others, and political behaviors like voting or influencing environmental policy.

Collective Environmental Action (Chapter 5)

Environmental actions entail working collectively with others. They include citizenship behaviors, such as engaging in protests and advocacy, as well as collective stewardship practices like volunteer tree planting or litter cleanups. Although environmental behaviors and collective action overlap, factors that predict individual behaviors may differ from those that predict collective action, which is why I devote a separate chapter to collective environmental action.

Knowledge and Thinking (Chapter 6)

The closer the knowledge and skills your audiences acquire are to the intended behaviors, the more likely those knowledge and skills are to lead to that behavior.

Generalized environmental knowledge is not likely to lead to behavior change or action, whereas action-related and effectiveness knowledge and systems thinking show greater promise.

Values, Beliefs, and Attitudes (Chapter 7)

Values are broad principles guiding what we do in life, whereas environmental beliefs have relatively little influence on our behaviors. Attitudes toward specific behaviors are more likely to influence behaviors than are general environmental attitudes. However, environmental attitudes can be hard to change, especially among adults. The environmental sociologist Thomas Heberlein (2012) compares attitudes to strong river currents and suggests that rather than try to change attitudes, we should learn to navigate them.

Nature Connectedness (Chapter 8)

People who feel a strong connection to nature are motivated to take action to protect it. Nature connectedness also contributes to emotional health and psychological resilience.

Sense of Place (Chapter 9)

Just as we can feel connected to nature, we can form attachments to a place. We associate certain meanings with places where we have lived, we depend on specific places for recreation and well-being, and we may even form an identity based on the places we know.

Efficacy (Chapter 10)

Our beliefs about whether our actions can achieve our individual and collective goals—that is, our personal, collective, political, and civic efficacy—determine the goals we set, the actions we take, and how persistent we are in trying to achieve our goals.

Identity (Chapter 11)

Identity is about the labels we give to ourselves, the groups we belong to, and how we distinguish ourselves from others. Although we often think of identity politics—appealing to particular ethnic, social, or religious groups—we also develop environmental identities, which influence our environmental behaviors.

Norms (Chapter 12)

Just as we can have individual and collective efficacy and individual and shared identities, we have personal and social norms. Personal norms are our expectations for our own behaviors, and influence our behaviors through making us feel guilty or giving us a sense of pride. Our perceptions of what most people *actually do* in a particular situation (descriptive social norms) or of *what more and more people are doing* (trending social norms) impact our behaviors more than what we are told we *should do* (injunctive social norms).

Social Capital (Chapter 13)

Social capital includes trust and social connections. When we trust and feel connected to others, we are more likely to work together for the common good—including stewarding our shared environmental resources. Environmental education programs where participants depend on each other to address a physical or other challenge can foster trust and social connections.

Positive Youth Development (Chapter 14)

Positive youth development is about acquiring assets important to success in life. Many of these same assets—efficacy, social connections, trusting others, and civic participation—also enable youth to engage in environmental behaviors and collective action. A focus on positive youth development allows environmental educators to partner with youth and community development professionals who view environmental education as a means for youth to acquire life skills.

Health and Well-Being (Chapter 15)

Similar to positive youth development, a focus on health and well-being enables environmental educators to develop ties with organizations that prioritize social issues. Whereas concerns about environmental quality are often described in opposition to concerns about human well-being, the evidence is clear that spending time in nature contributes to health and happiness. Importantly, people are motivated by finding meaning in life; spending time in nature, as well as stewarding and restoring nature, gives our lives meaning. In short, nature and health and well-being work hand in hand.

Resilience (Conclusion)

Similar to environmental quality, resilience is an ultimate outcome for environmental education. Similar to sustainable development, resilience integrates social alongside environmental concerns. It refers to how individuals and communities respond to ongoing change and catastrophic disruption by adapting and transforming. Thus resilience is particularly relevant in an era of rampant and rapid social and environmental change. Education programs embedded in civic ecology practices, such as community gardening, tree planting, and litter cleanups, can foster psychological, social, and ecological resilience as participants benefit from the healing power of nature, social connections, and seeing the positive results of their collective efforts on the environment and community.

Appendix

The appendix includes survey tools for measuring environmental education outcomes covered in the book chapters.

Environmental Quality: The Ultimate Outcome of Environmental Education?

Environmental quality, including sustainability and climate change mitigation, are often the ultimate outcomes of environmental education. Considering these ultimate outcomes shifts the focus from what participants think, feel, and do to the environmental impacts of their actions. Psychological and social factors like efficacy, identities, and norms are intermediate outcomes in pathways to behaviors and collective action, which in turn lead to environmental quality. Environmental quality is necessary for humans to thrive and is also important because of nature's intrinsic value. However, because of the impact humans already have had on the environment, for example on our climate, we are forced to adapt, preferably in ways that also mitigate future negative impacts. Thus, in addition to environmental quality, we discuss climate adaptation in chapter 3 on environmental quality. Resilience, another ultimate outcome that recognizes the need to adapt and transform in light of ongoing change and incorporates social alongside environmental factors, is discussed in the concluding chapter.

Some may object to a primary focus on environmental quality outcomes of environmental education. They ask whether this approach is too "instrumental"; that is, education becomes a tool *for the environment* rather than *for youth* to develop their competencies or realize their potential. Holders of this view might object to environmental education action programs whose goal is to increase ecosystem services or reduce greenhouse gases, for example by engaging youth in constructing rain gardens or joining a climate protest (Dietz et al.

2004). Yet many environmental action programs, such as those where youth work in community gardens or intern with local government officials, integrate environmental and youth development outcomes (Jensen and Schnack 1997; Wals et al. 2008; Schusler et al. 2009; Delia and Krasny 2018). Further, work on sustainability and resilience increasingly recognizes the intricate and inextricable ties between social and environmental outcomes. In short, programs don't need to be either for the environment or for youth. In fact, engaging successfully in civic life, including volunteer environmental actions, may contribute to critical thinking (see chapter 6), self-efficacy (chapter 10), social capital (chapter 13), youth development (chapter 14), and well-being (chapter 15), as well as lead to environmental outcomes.

Knowledge-Attitudes-Behavior: A Debunked Theory of Change That Persists

> **Here in Europe around 75% of the population believes that climate change is a very serious global problem. Europeans classify climate change as the third most serious problem in the world (after hunger and terrorism) so there is not much need to convince people about the existence and threat of climate change. However, very few people actually change their behaviour. The step from "knowing" to "doing" seems to be the hardest one.**
>
> (participant in Cornell online course, 2018)

In 1977, 265 delegates from sixty-eight countries gathered with representatives from the United Nations in Tbilisi, Georgia, USSR. There they issued a call to action: environmental education should help address environmental problems (UNESCO 1978). Their definition of environmental education seemed logical at the time:

> Environmental education is a learning process that increases people's knowledge and awareness about the environment and its associated challenges, develops the necessary skills and expertise to address the challenges, and fosters attitudes, motivations, and commitments to make informed decisions and take responsible action. (NAAEE, n.d.)

We can visualize the Tbilisi Declaration theory of change as follows: environmental education activities create the knowledge, skills, and awareness needed to address environmental challenges and foster attitudes, motivations, and commitments, which lead audiences to make informed decisions and to take action. This is simplified as the Knowledge-Attitudes-Behavior theory of change.

> The traditional thinking in the field of environmental education has been that we can change behavior by making human beings more knowledgeable about the environment and its associated issues. This thinking has largely been linked to the assumption that if we make human beings more knowledgeable, they will, in turn, become more aware of the environment and its problems and, thus, be more motivated to act toward the environment in more responsible ways. Other traditional thinking has linked knowledge to attitudes and attitudes to behavior. An early and widely accepted model for EE [environmental education] has been described in the following manner: "increased knowledge leads to favorable attitudes . . . which in turn lead to action promoting better environmental quality." (Hungerford and Volk 1990, 258)

Environmental education scholars Hungerford and Volk go on to warn that "most educators firmly believe that, if we teach learners about something, behavior can be modified. In some cases, perhaps, this is true. However, in educating for generalizeable [*sic*] responsible environmental behavior, the evidence is to the contrary" (267). Twenty years later, environmental education researcher Joe Heimlich reinforced this warning, lamenting the "stickiness" of the knowledge-attitudes-behavior paradigm.

> No criticism of theoretical weakness [of environmental education] would be complete without the acknowledgement of the old "knowledge to attitude to behavior" or "attitude to knowledge to behavior" claims many environmental educators still hold to be true. There is not much consensus regarding how attitudes might affect and predict environmental behavior. . . . Even so, myriad educators and scientists continue to believe if people just know enough, they'll change. Or if they feel a certain way, they'll act differently. (Heimlich 2010, 183–184)

Environmental education has experienced a lot of changes since the Tbilisi meetings in 1977. Multiple practices have split off from the Tbilisi approach to environmental education. Perhaps the most important of these is Education for Sustainable Development, which emerged with UNESCO support in the early 1990s. Its goal was to broaden environmental education to encompass social and economic justice, or "to empower and equip current and future generations to meet their needs using a balanced and integrated approach to the economic, social and environmental dimensions of sustainable development" (Leicht et al. 2018, 7). The emphasis on social justice is an invaluable contribution and has been followed by similar efforts such as community and youth development approaches to environmental education. Yet as recently as 2018, a key UNESCO publication states,

> Education for Sustainable Development (ESD) is commonly understood as education that encourages changes in knowledge, skills, values and attitudes to enable a more sustainable and just society for all. . . . The concept of ESD was born from the need for education to address the growing environmental challenges facing the planet. In order to do this, education must change to provide the knowledge, skills, values and attitudes that empower learners to contribute to sustainable development. (Leicht et al. 2018, 7)

I advocate that we liberate ourselves from a narrow definition of education as transmission of knowledge and skills, or even attitudes. Instead, I envision environmental education more broadly as "all forms of formal, non-formal and informal education and training that equip individuals and institutions in the public, private and community sectors *to effectively respond to pressing environmental challenges*" (Wals and Benavot 2017, 405, italics added). In short, if we start with the big goal of how to address environmental challenges, we can broaden our tent to encompass the cognitive and affective capacities—whether they be action-related knowledge, nature connectedness, sense of place, efficacy, identity, norms, or social capital—that studies have demonstrated influence environmental behaviors and collective actions.

Node in the Network

> **To be most impactful, education and lifelong learning should be part of an integrated approach that also includes changes in governance, legislation, research, financing and regulation towards greater environmental sustainability.**
>
> (Wals and Benavot 2017, 405)

In their book, *The Failure of Environmental Education*, Saylan and Blumstein (2011) claim that the environmental crisis is evidence that environmental education has failed. But no one ever suggested that environmental education alone can change the world. We might describe environmental education as "one tool in the toolbox"—or "one node in the network"—of efforts to improve the environment. Environmental education works alongside laws and regulations, research, the private sector, and civil society advocacy and voluntarism to make a difference.

Barry Commoner, the renowned scientist and activist whom *Time* magazine dubbed the "Paul Revere of Ecology," famously said, "The first law of ecology is that everything is related to everything else" (C250 2004). A century earlier, John Muir pointed out, "When we try to pick out anything by itself we find that it is bound fast by a thousand invisible cords that cannot be broken, to everything in the universe" (Sierra Club 2018).

Environmental educators are familiar with the connections described by Commoner and Muir and in fact incorporate such "systems thinking" into their programs. But we can also apply Muir's web of invisible cords to elucidate how environmental education programs are connected to other forces working for environmental and educational change. For example, environmental education connects to laws and regulations when program participants identify and research a problem and work with local officials to implement new policies. Environmental education connects with research through citizen science and other types of data collection efforts. It connects to the private sector through corporate social responsibility initiatives, such as community cleanups or support of renewable energy. And environmental education connects to community organizations, which provide internships for youth, partner with nature centers to conduct Earth Day festivals, and join in activities ranging from hands-on stewardship to environmental advocacy. The North American Association for Environmental Education recognizes these connections, stating, "Environmental education is a key tool in expanding the constituency for the environmental movement and creating healthier and more civically-engaged communities" (NAAEE, n.d.).

But environmental education is not just part of a network of initiatives and organizations. It is also part of a network of solutions. Three broad "fixes" categorize our efforts to address environmental problems. Technological fixes involve changing the environment directly (for example by installing solar panels). Structural fixes entail changing laws and regulations. Most would describe environmental education as a cognitive fix, which relies on people changing in response to new information (Heberlein 2012). However, here again a web of connections more accurately describes the situation. Environmental education has affective as well as cognitive outcomes. Youth environmental action programs may directly change the environment, for example through restoring habitat or reducing greenhouse gases. They also can influence policy. It is critical to take into account the technological, political, and other structural barriers that limit what we can do; environmental education is not a panacea. But environmental education is not only a cognitive fix. It can also be part of technical and policy solutions to environmental problems.

In short, a network of government, private, and civil society organizations work in a web toward a greener environment, and environmental education organizations are nodes in this network. Online social media accelerates this trend toward "networked" environmental governance—or policy formation through both formal government and civil society organizations (Bennett and Segerberg 2013; Connolly et al. 2014). It is this governance network that can address policy changes and other structural barriers that constrain the ability of environmental education to reach its goals. At the same time, by leveraging individual partnerships and by being part of a broad network of organizations, environmental education can help accomplish what it is unable to accomplish alone.

This book poses the question, What if instead of starting with knowledge and attitudes, we begin with factors that research has demonstrated predict environmental behaviors and collective action? What if we draw on research in environmental psychology, sociology, economics, and political science to inform our programs? In answering these questions, I ask the reader, in addition to focusing on better ways to teach, to consider environmental education as a broader cultural and social force that includes knowledge and attitudes, but also efficacy, identity, norms, connections, and trust. And just as environmental education can benefit from connecting with research across multiple disciplines, it should not try to tackle the environmental crisis alone. Through forming partnerships with the network of government, private, and civil society organizations working toward the public good, we collectively have the capacity to achieve our common goals.

Part I
GETTING STARTED

1

THEORY OF CHANGE

A theory of change offers a picture of important destinations and guides you on what to look for on the journey to ensure you are on the right pathway.

(Reisman and Gienapp 2004, 1)

Highlights

- A theory of change is a diagram and narrative explaining how program activities lead to program outcomes.
- Creating a theory of change can help environmental educators think critically about planning a new program, reflect on how to improve an existing program, identify targets for evaluation, and communicate about a program.
- The reflective and collaborative process of creating a theory of change can lead to new insights and new opportunities for working with stakeholders.
- Your theory of change should include a diagram and narrative that describe environmental or other ultimate outcomes, behavior or collective action outcomes that lead to ultimate outcomes, intermediate outcomes that lead to behavior change and action, and activities that lead to intermediate outcomes, as well as assumptions and context that may influence desired outcomes.

What Is a Theory of Change?

Although you may not articulate it out loud, somewhere in your mind you likely have a theory of change about how your environmental education program makes a difference to participants, your community, and the environment. For example, you may live in a coastal town, and you want your rural participants to support local greenhouse gas mitigation policies. You think the pathway to get there is for them to develop trust with government officials and university climate scientists. So you organize an activity—a volunteer day where youth, community members, county officials, and scientists work side by side to install oyster reefs along the shoreline. You considered an alternative pathway—an evening lecture by a climate scientist—but knowing that knowledge does not generally lead to action, especially for community members who may be climate science skeptics, you decide that informal sharing and learning while volunteering may be a more effective pathway to meet your goals.

Articulating your theory of change helps you to think broadly about your big outcomes, such as improved environmental quality. You then define changes in behaviors or collective actions that will lead to your big outcome, and identify the intermediate outcomes—like trust and environmental identity—that are needed to effect those changes. Once you have defined your environmental behavior or collective action, and intermediate outcomes, you home in on what program activities will most likely lead to your intermediate outcomes. By diagramming your theory of change, you create a visual pathway to reach your intermediate and behavior/action outcomes; this theory is based on your own experience, the experiences of your colleagues and community members, your reflections, and the results of research and evaluation. A complete theory of change also includes a short narrative where you explain the context, assumptions, and reasoning behind your pathway diagram.

You may be familiar with logic models, which have some similarities with theories of change. A logic model is a diagram of a program's components, including its inputs (e.g., funding, expertise); activities (e.g., curriculum development); outputs (e.g., a curriculum and teacher training); intermediate outcomes (e.g., teachers' knowledge gain); and long-term outcomes (e.g., students of a teacher who has gone through your training will increase their scores on the state science exam). Similar to a theory of change, a logic model allows you to see if your outcomes are in sync with your program activities. However, a theory of change goes one step further—it forces you to reflect on *why* you predict certain activities will lead to desired outcomes. You can think of a logic model as a *description* of a program and its outcomes, whereas a theory of change is an *explanation* of the pathways to reach a program's outcomes (Clark and Anderson 2004). Through the process of explaining, we also engage in critical reflection about our assumptions. In short, a logic model is used to make sure you have all the pieces in place for your program and is often used by funders to evaluate project proposals. A theory of change encourages deeper reflection—including rethinking and adapting programs based on new information.

Why Is a Theory of Change Important?

> **Failure in reaching goals is almost guaranteed in the absence of a clearly developed model of change. Failures in the context of a Theory of Change can be opportunities to learn from the experience, recalibrate, and return to the field with more effective interventions.**
>
> (Taplin et al. 2013, 7)

The iterative process of constructing a theory of change with colleagues and other stakeholders is at least as important as the final product.

- The process of constructing a theory of change allows staff and stakeholders to learn from each other's experience and from related research and to engage in dialogue that challenges their assumptions. Questioning established ways of thinking is particularly important in environmental education, where we often default to the "knowledge-attitudes-behavior" theory of change that has been debunked by research over the last forty years (Hungerford and Volk 1990; Kollmuss and Agyeman 2002; Heimlich 2010).
- Theories of change help to define which outcomes should be the focus of our program evaluations. For example, do we want to measure changes in trust among program participants, a new collective action, or a change in environmental quality? Although not all environmental education programs have the resources to systematically assess different levels of outcomes, diagramming and describing the pathways in your theory of change can at a minimum focus your observations and reflections on what is working, what is not working as you thought it would, and what adjustments you might make to your program activities and theory of change (Connell and Kubisch 1998; Taplin et al. 2013).
- When constructed collaboratively with colleagues and even partner organizations, a theory of change can build relationships among a range of stakeholders (Taplin and Clark 2012; Taplin et al. 2013).
- A theory of change can be used to communicate about your program to stakeholders and funders. You might consider simplifying your theory of change diagram into a logic model, which may be easier for a funder or other stakeholder to grasp (Taplin et al. 2013).

In short, constructing a theory of change is important in program planning and evaluation, communicating about program goals and activities, establishing partnerships, and, perhaps most important, helping educators to critically question assumptions and, when necessary, replace them with research- and experience-based strategies.

This Book Has a Theory of Change

You might ask, what is the theory of change of this book? My ultimate outcome, as with many environmental education initiatives, is to improve environmental quality and build social-ecological systems resilience (see chapters 4 and the conclusion). To reach that ultimate outcome, I believe environmental educators can play a key role.

For environmental educators to play a role in improving environmental quality and building resilience, I believe we need to reflect on—and, where needed, enhance—our practice, which is my behavioral outcome. In particular, as environmental educators, we need to broaden the range of approaches we consciously use to influence behavior. We often focus on knowledge and attitudes, but our programs can influence multiple intermediate outcomes that have been shown to impact individual behaviors and collective action—including efficacy, norms, identity, and social capital. I include these intermediate outcomes because my review of the research and my experience lead me to believe that, compared with instilling knowledge or trying to change attitudes, these intermediate outcomes are oftentimes more effective in changing behaviors and actions.

Yet, ironically, a book like this one is an attempt to build knowledge among readers. I believe, however, that unlike many environmental education audiences who may not have an environmental identity or hold environmental values, the readers of this book—practicing or aspiring environmental educators—are eager for new knowledge, in particular action-related and effectiveness knowledge (chapter 6) to help them reach their goals. At the same time, environmental educators are often isolated and may not be supported by their organization as they try to adopt new practices. I also believe that there is no one answer to solving the environmental crisis, and thus that the experience of a broad group of educators and researchers is critical to helping us try different approaches and to learn based on the results of our efforts. For these reasons my theory of change goes beyond this book. It also includes forming social networks to help environmental educators learn from and support each other as they enhance their practices. Thus, in addition to this book itself, my theory of change includes creating opportunities for sharing knowledge, practice, and resources, and for forming social connections through social media, online courses, and face-to-face workshops (figure 1.1; see Civic Ecology Lab 2019).

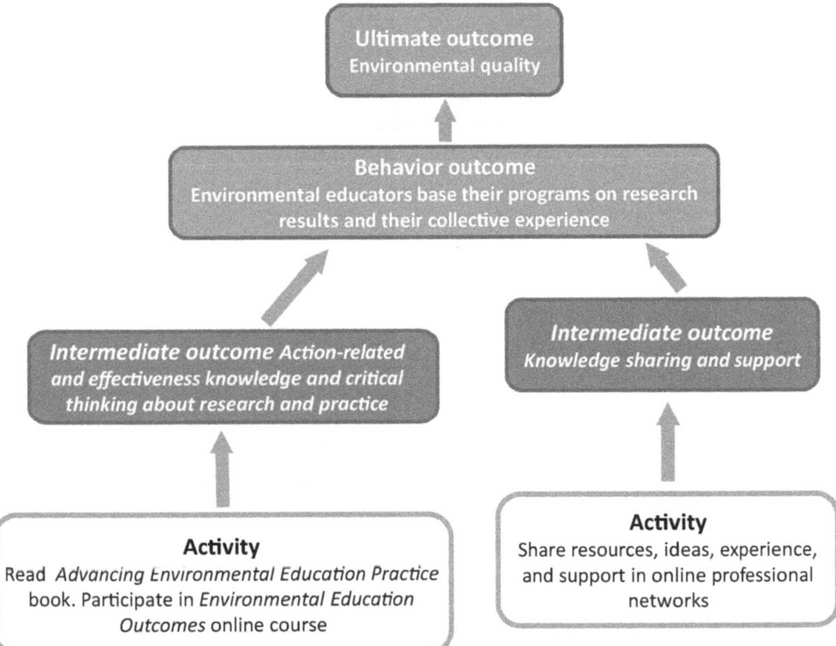

FIGURE 1.1. Theory of change diagram for this book and related Civic Ecology Lab activities

Constructing a Theory of Change

> **Developing a program's theory of change can, thus, allow researchers and practitioners to look inside the "black-box" and examine the mechanisms that lead to desired changes and outcomes.**
>
> (Burbaugh et al. 2017, 194)

Before embarking on your theory of change, you might ask yourself: why bother? For what purpose will I invest the time and energy to develop a theory of change? Questions to guide your thinking about the purpose of a theory of change include the following: Do you want to reflect on your own assumptions about a program you are responsible for? Do you want to join with colleagues to develop a new program? Would you like to identify areas for assessment in an existing or new program? Or maybe you want to collaborate with government agencies, businesses, and nonprofit organizations to develop a coordinated theory of change that will inform environmental education in your state?

Note that theories of change can be constructed individually or as a collaborative exercise, and can be used to reflect on, improve, plan, and communicate about your program or a broader initiative. Steps in creating a theory of change include (1) articulating your ultimate outcome, (2) articulating your behavior or collective action outcome, (3) articulating intermediate outcomes likely to lead to the behavior or collective action, (4) identifying activities to achieve your intermediate outcomes, (5) considering the context, (6) constructing a narrative, and (7) reflecting and revising.

1. Ultimate Outcome

First, articulate your ultimate outcome. Ultimate outcomes are the conditions that will change—like reduction in greenhouse gas emissions or increase in pollinator populations—as a result of actions taken by program participants (Burbaugh et al. 2017). In reality, this big outcome is likely to be achieved in partnership with multiple organizations, government agencies, and the private sector, but it is still important to keep in mind one's ultimate goal because it guides your behavior/action and intermediate outcomes, as well as your program activities. Not all environmental education programs focus on environmental quality as their ultimate outcome. For example, a science-based program may aim to increase science literacy and ultimately to enhance the nation's technology competitiveness, and a youth development program that incorporates environmental stewardship may be part of efforts to improve community well-being.

Ultimate outcome:
Environmental quality (reduced greenhouse gases)

2. Behavior/Action Outcome

Next think about what your program participants need to do to realize your ultimate outcome. Let's say that your ultimate outcome is slowing climate change by reducing greenhouse gases. Can participants change their individual behaviors to reduce greenhouse gases? Can participants work together to take collective action? Although multiple pathways are possible, based on recent reading and your knowledge about state tax incentives for installing solar, you decide your pathway for achieving the ultimate outcome is for your city to install a community solar farm. By reducing fossil fuel consumption of multiple households, community solar will achieve your ultimate outcome of reducing greenhouse gases. But for that to happen, the most promising pathway seems to be for your program participants to work together with community groups to influence local renewable energy policy. Collective action to influence policy is then your behavior/action outcome. In short, the behavior/action outcome is a second-level outcome that leads to your ultimate (top-level) outcome.

Collective action outcome: Advocate for renewable energy policy

3. Intermediate Outcomes

The third step is to think about intermediate outcomes, which define your pathway to achieve your behavior/action outcome. How do you build the capacity of program participants to advocate for your local government to approve a community solar project? Perhaps you have read the literature on political efficacy, which suggests that people who have had positive experiences changing policy will acquire a "can do" attitude and be more likely to participate in additional policy actions (see chapter 10). Your pathway is beginning to take shape. Political efficacy leads to collective action to influence energy policy to reduce greenhouse gases.

Note that you may have several pathways to reach your behavior/action goal, each with its own intermediate goals. In addition to political efficacy, you remember that the research on collective environmental action suggests that for people to work together, they need to develop some sort of trust or build "social capital" (see chapter 13). You decide to include both political efficacy and social capital as intermediate outcomes or pathways to collective action.

Intermediate outcomes: Political efficacy and social capital

4. Activities

What activities foster political efficacy and social capital? In planning your activities, you can bring in the research as well as your own experience of what you have seen work in similar situations. You may also want to get input from colleagues and community and family members.

The research on efficacy suggests that people who achieve initial success through "mastery experiences," and who have positive role models and supportive social interactions, are likely to develop a sense of self- and political efficacy (Bandura 1977; Beaumont 2010). Research suggests that social capital can be built through social, recreational, and challenging activities in which participants build trust and connections (Krasny, Kalbacker, et al. 2013).

Maybe your colleagues led an environmental action program, in which students attempted to reverse a local policy regarding new highway construction. Unfortunately, the students got to the point of presenting their argument to the transportation department, but the transportation department failed to act. Upon reflecting on that experience, you tease out several lessons learned. Perhaps the project was too ambitious—your colleagues failed to account for the influence of more powerful businesses. Perhaps there were things outside their control—structural factors like federal highway dollars—that were working against the students. And maybe your colleagues could have teamed up with other organizations like homeowners and environmental groups, rather than try to go it alone. You have seen another environmental action program in your school achieve its goals and decide to talk with the teachers leading that effort to determine what they think enabled success.

After drawing on research and experience, you are ready to propose program activities to reach your intermediate outcomes. You decide to start with activities to build social capital. You organize a volunteer activity, where participants work with community members to harvest food at a church garden and deliver it to a food kitchen. In working together, program participants and community members build trust and social connections.

Second, you plan a small project that serves as a mastery experience to build participants' self- and political efficacy. You are aware that school cafeteria personnel want to reduce waste, but they don't know how. Your students research what other schools have done and make recommendations to the cafeteria. Since the cafeteria has already indicated its interest in the waste issue, students are likely to be able to influence school policy, and thus have a mastery experience. At the same time, they will be going through the steps used to build action skills, including research, critical thinking, and communication (Earth Force, n.d.).

Now that your students have acquired a degree of trust and social connections and efficacy, you are ready to engage them in research, critical thinking, and communication activities that focus on advocating for a community solar farm (figure 1.2).

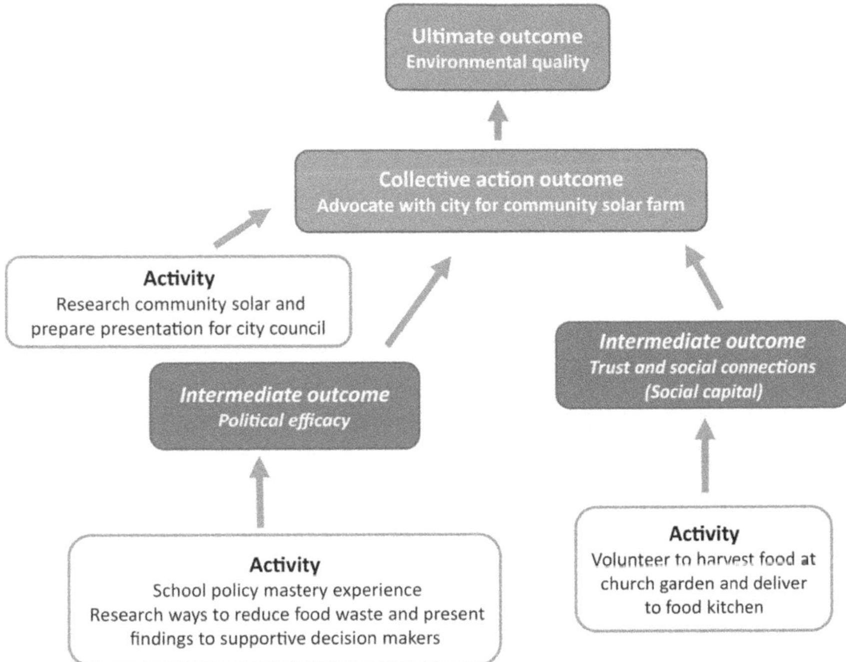

FIGURE 1.2. Theory of change diagram for environmental education program

5. Context

As you create your theory of change, keep in mind factors outside your program that might impact its success—in other words, consider the context. For example, if you are trying to influence local solar power implementation, are there new state or federal incentives for community solar coming online? Or is the recent implementation of tariffs on foreign-made solar panels likely to increase their costs? Is a key policy maker about to leave his position and be replaced with someone favorable to your initiative? And what partners might you engage in the program activities to make those activities more meaningful, or even in discussing your theory of change to make it more robust? Outside forces both constrain and provide opportunities for your program (Connell and Kubisch 1998; Taplin et al. 2013).

6. Narrative

Once you have completed your theory of change diagram with activities and three levels of outcomes, you will want to write a short narrative summarizing the reasoning behind your proposed pathway(s). Your narrative should describe in one or two paragraphs the ways in which the activities lead to the intermediate, behavior/action, and ultimate outcomes. It should also include two types of reasoning: (1) why intermediate outcomes lead to the behavior/action outcome (If this occurs, then . . .), and (2) why activities lead to intermediate and behavior/action outcomes (If we do this, then . . .). Also, describe contextual factors that might influence your ability to reach your outcomes. If you are using your theory of change to inform your evaluation, you may also want to include the indicators you will use to measure the various outcomes.

7. Reflecting and Refining

> **Theory of Change allows proponents and stakeholders the means to continually challenge their assumptions and, in doing so, refine and sharpen their strategies for greater success.**
>
> (Taplin et al. 2013, 8)

Whether or not you achieve your ultimate and behavior/action outcomes, it is important to reflect on your proposed pathway. Constructing a theory of change involves not just creating a diagram and narrative but continually challenging your assumptions and, if needed, refining your program activities and proposed outcomes. You might ask, What is my evidence that the volunteer activity fostered social capital and that social capital in turn led my program participants to work together for change? What if they didn't build trust through the volunteer activity? Should I conduct multiple activities before suggesting they work together on a policy issue, or should I abandon my approach and revise my theory of change? Or perhaps I should revise my collective action outcome?

If you're like me, it's difficult to admit that something might not be working as intended; we all get invested in our program activities. We also all have some underlying theory about how we can make a difference, and we can benefit by reflecting on and tweaking our plans and activities as new information becomes available. Articulating our theory of change forces us to pay careful attention to the ways in which our participants are using the opportunities we afford them.

Sometimes program participants can help open up that black box of what actually happened in a program and shed light on your theory of change. Graduates of an agricultural leadership program in Virginia reflected on what they had learned, identified salient behavior outcomes, and diagramed the connections between activities and these outcomes (Burbaugh et al. 2017). Only after they had identified changes in their leadership practice and the program activities that led to these changes did the intermediate outcomes or theory of change pathways become obvious. In short, the program participants collaboratively created a visual model of how the program had developed their leadership capacity, thus providing the program leaders with new insights.

In the above example, the program leaders wanted to enrich their understanding of how the activities had led to outcomes from past participants' perspectives. In traditional applications of theory of change as a planning tool, educators propose pathways and intermediate outcomes before deciding on activities to achieve those outcomes. However, you might also want to consider a pathway mapping activity conducted by past program participants to gain insight into your program and theory of change.

Using Your Theory of Change for Evaluation

A theory of change can be used to identify outcomes that can be evaluated. With the help of an evaluator, you can then define indicators for your outcomes (see chapter 2). Indicators can be quantitative, such as the number of people working to change a policy, or descriptive, such as documenting a new policy. To estimate the impact of a community solar farm on greenhouse gas reduction, you might work with a local utility to gather information on the kilowatts of solar power produced and its equivalent in the volume of CO_2 or methane gas reduced. For intermediate outcomes, like political efficacy or social capital, you can use existing surveys (see appendix). A qualitative evaluation using interviews and observations aids in understanding how and why a program works—which is valuable information as you go about adapting your program to reflect your current understanding of outcomes and how they are achieved (Connell and Kubisch 1998).

If you find that you have not reached your desired outcomes, an evaluator might help you identify what changes need to be made. For example, perhaps the implementation of your activities was not done in the most effective way. Maybe your assumptions about which activities lead to particular outcomes were not valid, or the situation changed mid-program, presenting new structural barriers. Maybe your theory needed to be expanded to take into account additional intermediate outcomes as well as outside factors that you are not able to control (Taplin et al. 2013).

In the end, you want your theory of change to be *plausible*—it reflects what can happen to the best of your knowledge. You want it to be *doable*—you can implement the activities given the resources available to you. And you want it to be *testable*—you can either evaluate it formally or you can use it to guide your own observations and reflections. Because environmental educators work in situations where information about what activities might lead to particular intermediate outcomes, and what intermediate outcomes lead to behavior/action outcomes, is constantly changing based on new research, be open to revising your theory of change as new information becomes available (Connell and Kubisch 1998).

2

EVALUATION

[I believe there is] an inevitable shift from evaluation being used to "confirm the quality" of an educational program, to being a tool that examines the efficacy of education practices leading to conservation outcomes.

(Heimlich 2010, 184)

Highlights

- Whereas environmental educators sometimes conceive of evaluation as an unwelcome obligation or an opportunity to prove their success, a "learning through evaluation" culture focuses on opportunities for learning through formal assessments and through ongoing observations and reflections.
- Evaluation presents an opportunity to revisit our theories of change, initially to specify outcomes to evaluate and later to adjust our activities, outcomes, and theory of change based on findings.
- Educators can partner with professional evaluators to conduct rigorous assessments of their program outcomes, often using pre- and post- surveys and control groups.
- Educators can also use "Most Significant Change" approaches to learn about impacts on participants and embedded assessment tools that serve as both learning activities and outcome measures.

As an environmental education researcher, I sometimes get asked by my environmental education colleagues to evaluate their programs. I see that the environmental educator is passionate about her work, and desperate to prove its impact to the wider world. While truly admiring the work of the educator, I feel a little uncomfortable with the request. In the back of my mind is the nagging question: What if the evaluation showed that the educator's work was not having the intended impact on most participants? And is the educator's intended impact stated outright and supported by a theory of change? These two questions reflect the two big messages of this chapter: First, no one program achieves or fails totally in reaching its intended outcomes. However, when we are open-minded, a formal evaluation and even our everyday observations and reflections can unlock ideas about how we might work with our participants, colleagues, community, and funders to achieve common goals. Second, evaluations and theory of change go hand in hand—oftentimes what is perceived as failure is in reality a program that is based on a faulty theory of change.

To reinforce and expand on these points, in this chapter I first discuss multiple reasons why environmental educators might want to evaluate their programs. Then I cover three broad approaches to evaluation, a generalized evaluation process, and ethical considerations in evaluating participant outcomes. This chapter provides a broad overview of evaluation issues and approaches; several resources are listed at the end for readers needing a more detailed step-by-step evaluation guide. Further, tools for assessing outcomes are included at the end of the outcomes chapters in parts 2 and 3, and in the appendix.

In writing this chapter, I adhere to a broad definition of "evaluation" and "evaluators." I include highly trained social scientists who conduct rigorous evaluations that meet research standards. Partnering with a competent researcher can provide valuable information, especially given the challenges of environmental education evaluation—from deciding which behaviors to measure to designing rigorous measurement tools for myriad possible outcomes consistent with theories of change (Heimlich 2010; Frantz and Mayer 2014). My definition of evaluators also includes environmental educators, who while perhaps not trained in evaluation, interact with program participants on a daily basis and thus can share insights, observations, and reflections that provide valuable lessons about outcomes and pathways to reach them.

Why Evaluate?

> **One of the potentially valuable roles for evaluation in environmental education settings is to build the capacity of both the individual educator and all the educators across the organization to improve programs through data-driven decisions.**
>
> (Heimlich 2010, 184)

Evaluation takes time and resources. So before launching into an evaluation, it is important to ask why. You may want to know how well your activities achieve their intended outcomes for participants. But other reasons for evaluating are possible—evaluations can spur discussions among colleagues and between education organizations and funders, and the results can be used to improve programs. Conducting evaluations can also contribute to professional and organizational development (Ernst et al. 2009; Greene 2010). Paralleling environmental education's emphasis on experiential learning, I view evaluation as part of the experience of being an environmental educator and subscribe to a learning-through-evaluating process.

The challenge for all of us doing evaluations is to be open-minded about the results. What if you discover through an evaluation that only a small fraction of your program participants are reaching your intended outcomes? Whereas negative results can be disappointing, the trick is to learn from rather than discount them. Treat any results—positive and negative—as a learning opportunity. Which activities seem to be achieving outcomes for which participants, and which activities need to be adapted or abandoned? Even without a formal evaluation, you can observe, talk to program participants, discuss issues with colleagues, and use what you know about environmental education to continually revise your program. Regardless of the results, you can also revisit your theory of change (see chapter 1). Does your theory reflect what the research shows about how people learn; how attitudes, values, identity, norms, social capital, and other intermediate outcomes are developed and maintained; how environmental behaviors change; and what factors are important to collective action? The outcomes chapters in part 3 of this book summarize this research and are intended to help you articulate and, where necessary, question your theory of change. The information is also intended to help you abandon outdated theories of change (e.g., theories that subscribe to disproven truisms, such as imparting environmental knowledge will change environmental behaviors). Through the information in this book and your formal evaluations, as well as your observations, reflections, and discussions with colleagues, you can replace a theory that may not be supported by research with a new theory that includes realistic outcomes and research-based pathways for getting there.

The truth is that nearly all evaluations yield mixed results—some program participants may achieve the desired outcomes, others will not. Some may engage in only one environmental behavior, others will be motivated by your program to engage in additional "spillover" environmental behaviors (see chapter 4). Use evaluation results as a chance to reflect, discuss with colleagues, read about what has worked in other programs, and adjust your theory of change and program accordingly.

Learning through Evaluation

How might you respond if the evidence does not clearly support your theory of change, or your belief that your program activities achieve your intended outcomes? Likely you need to reassess your program activities, alter your intended outcomes, or consider the appropriateness of the activities for your audience. This "adaptive programming" is normal—educators do it on a lesser scale as they judge their audiences' reactions and adjust their actions every day. Helping participants in an environmental education program learn new material, develop self-efficacy, become engaged in civic life, and change their behaviors is an extremely complex process, and there is never one pathway to get there. It is to be expected that we are continually adapting our approach based on our goals, our observations, the latest research, and more formal evaluations. In fact, continually reassessing how we are doing can be a catalyst for sharing concerns and observations with colleagues, and for rethinking and trying new things—all of which, while challenging, can be a rewarding part of our jobs.

Of course, the reality is that educators are often under pressure to provide evidence of their success for funders, bosses, colleagues, parents, and other stakeholders. However, funders and other stakeholders who have an investment in your program are likely interested in how you are constantly improving your program, in addition to your program's actual successes. Hopefully you can engage in an ongoing and open discussion with your funder about what is working and what you are doing to better meet your goals. You might want to agree with your funder on an evaluation strategy at the outset, so that there is transparency around your desire to engage in ongoing improvements; this also provides a bit of a buffer against the risk of your funders changing their priorities or expectations partway through a grant. Even when your program is working well and you can provide evidence of outcomes, funders may have their own ideas about what is important. Finally, you might want to talk with your funder about how to report data; some prefer hard statistics, others may respond more to stories, photos, and quotes.

Author Reflections

The Chinese Alibaba Foundation has provided funding for our Civic Ecology Lab online courses targeting environmental educators. We recently produced a report outlining the impact of one online course on participants' learning, their development of professional networks, and how they intend to apply what they learned to their environmental education programs. We used both summaries of outputs, such as how many people signed up for the course and how many completed it, and of outcomes, such as what participants felt they learned and their development of social networks. We also included quotes from participants indicating how they intended to apply what they learned in their work (Russ et al. 2017). Our Alibaba Foundation project officer appreciated the report but is under pressure from her boss to show that their funds are reaching larger numbers of people. We are making adjustments in our strategy accordingly, in part by posting more Chinese translations of our research summaries on popular Chinese social media. At the same time, we are careful to ensure that the new activities implemented to meet the funder's interests are consistent with our Civic Ecology Lab program goals.

In short, evaluations can be stressful, but most importantly, they can be viewed as an opportunity. They can show what you have achieved, and they can help in the ongoing process of program and professional development. They can provide impetus for reflecting on and revising one's theory of change, program activities, and even program audiences and goals. And they can help build the capacity of your organization to offer programs that have an impact on participants, the community, and the environment (Ernst et al. 2009). It is critical to define the type of evaluation that suits your professional, program, and organizational needs, and when working with outside evaluators to engage in a participatory process to realize the multiple benefits of evaluation.

Evaluation Approaches

> **Many educators who may be willing to concede that one trip to the zoo for third graders may not change a career path still stubbornly cling to the notion that educational programs that raise awareness and provide excellent experiences can bring a host of glorious changes.**
>
> (Monroe 2010, 195)

Evaluations can be formative or summative. Formative evaluations conducted during a program offer feedback to the educator about what is going well and what adjustments should be made. Summative evaluations measure final outcomes after a program has ended and can be used to inform funders and administrators about the worth of the program and help them make decisions about whether to continue it (Simmons 2004; Ernst et al. 2009; Monroe et al. 2016). However, many evaluations fall somewhere in between formative and summative because educators and funders want both to make improvements and document outcomes.

Needs assessment conducted before a program is sometimes considered a type of evaluation. It is used to understand participants' needs and inform the development of environmental education activities (Meichtry and Harrell 2002). An audience assessment is also conducted before a program to determine factors such as participant attitudes, values, and social-political identity (Monroe and Nelson 2004), which may be difficult to change in an environmental education program but can be taken into account in program planning. Rather than focus on changing strongly held attitudes, consider "navigating" your program participants' attitudes to achieve common goals (Heberlein 2012; see chapter 7). So, for example, when working in rural Virginia, my colleague Anne Armstrong used her interactions with local residents and published information on US climate change attitudes and cultural identity to plan community stewardship activities, like installing oyster reefs along the town's beach, that would appeal to all participants regardless of their attitudes and identities.

Below I outline three broad evaluation approaches: pre-/post-/control coupled with open-ended interviews, Most Significant Change, and embedded assessment. Many agree that pre-/post-/control is the gold standard of evaluation, as it can provide conclusive "proof" of the average outcomes of your program. However, pre-/post-/control designs generally use close-ended survey questions and thus may fail to capture unexpected outcomes. They also provide limited information about the process by which outcomes are achieved or fail to be realized. For this reason, evaluators often combine pre-/post- surveys with open-ended interviews. Most Significant Change evaluation helps program developers understand what outcomes are possible and how outcomes are achieved but does not provide a representative view of the typical participant outcome. "Embedded assessment" refers to program activities that also provide information on outcomes. Note that you can question evaluation partners about the approaches they use and choose an evaluator whose approach feels comfortable for your organization and addresses your funder's information needs.

For the outcomes chapters in parts 2 and 3 in this book, I provide example surveys and interview questions that could be used in pre-/post-/control evaluations (see also appendix), and embedded assessment activities where appropriate.

Most Significant Change evaluations use standard questions and thus are not covered in individual chapters.

Pre-/Post-/Control-Group Assessment and Open-Ended Interviews

In a pre- and post/control assessment, the evaluator measures targeted outcomes—for example, nature connectedness or social capital—before and after the program and calculates changes in the level of the outcome. Although any increases in outcomes revealed when comparing the pre- and post-surveys are likely to be a result of the program, the possibility exists that even without the program, a similar group of individuals would demonstrate similar changes. For example, if the evaluation showed increases in knowledge about climate, and climate change had been prominent in the news during the period the program was offered, this increased knowledge might have occurred anyway as a result of people reading, watching, or listening to the news. Another possibility is that increases in self-efficacy, social capital, or another outcome might have occurred as a result of any program of the same length—for example, a youth program focused on theater rather than on the environment. To boost confidence in evaluation results, evaluators sometimes include a control group composed of individuals with similar demographics to those in your program, but who have participated in a different type of program (e.g., theater) or not participated in any program. In both cases, evaluators would administer the pre- and post-test to program participants and to the control group. In some instances, you can use a delayed control design, in which you offer the environmental education program to the control group at a later date. This addresses the ethical quandary of using people as "controls" without giving them the benefit of your program.

If you are not able to take measures before a program starts, you can ask participants to reflect on any changes they experienced in their sense of place, self-efficacy, or other outcome using a retrospective evaluation design. This entails asking participants about the changes they experienced after the program ends. Although participants may find it difficult to recall how they thought or behaved prior to the program, using retrospective data has the advantage of asking participants specifically if the outcomes are a result of the environmental education program. In addition to examining your program's short-term impact immediately after it ends, you can conduct a follow-up evaluation several months or more after the activities to determine longer-term impacts.

The surveys used in pre-/post-/control evaluations can provide valuable quantitative information about what a program has achieved. However, the results are like a "black box"—they tell us a result but do not give us information about the process that led to any positive or negative outcomes. An educator or funder may look at puzzling results and want an explanation. For example, the pre-/post-test might show that girls in the program enhanced their systems knowledge, but the boys seemed to make greater gains in self-efficacy. To help untangle survey results, evaluators often conduct open-ended interviews with program participants before and after, or retrospectively after the program. For example, in interviews, boys might explain how bored they were with the activities meant to foster systems knowledge, but how much they enjoyed the challenges meant to build self-efficacy. This kind of in-depth information provides insight into what program elements lead to which outcomes, how hypothesized pathways to achieve objectives may need to be adapted for different audiences, as well as useful quotes for an evaluation report.

Pre-/post-surveys often include checkbox questions, in which respondents go through a list and check those that apply to them, for example a list of program activities they participated in. Surveys can also include Likert scale questions, in which participants rate their level of agreement or disagreement with a particular statement that reflects an outcome. For example, for an ecological place meaning outcome, participants might be asked to rate, on a five-point scale of strongly agree to strongly disagree, their level of agreement with the statement "My city is a place where I can see wildlife," or "My neighborhood is a place where I can enjoy taking walks in nature" (Kudryavtsev et al. 2012). Note that evaluators often use online survey tools, like Qualtrics or SurveyMonkey, which can generate summary graphs of the data.

Because results can be summarized numerically, surveys with checkbox and Likert scale questions are referred to as quantitative methods. They provide what funders often perceive as "hard data" showing evidence of your program outcomes. Qualitative methods, in contrast, include open-ended interviews, focus groups, observations, and analysis of participant-generated photos, videos, diaries, guided reflections, or concept maps. The evaluator codes the data using predetermined categories, such as those reflecting program goals, while paying attention to any new outcomes or insights that the evaluator did not anticipate. The results enable researchers and environmental educators to create rich descriptions and understand the complexities of their program (Patton 2002).

Most Significant Change

Using stories in evaluation (a) generated important understandings of the *interwoven character* of the program with its context, and

> **(b) provided windows of unique insight into participants' *lived experiences* of important program effects.**
>
> (Costantino and Greene 2003, 36, italics in original)

Most Significant Change, or MSC, captures participants' stories of the most important change they experienced during a program (Dart and Davies 2003). Evaluators ask participants a question such as, "During the last month, in your opinion, what was the most significant change that took place in the program?" Participants respond by writing or telling a story that includes details about the outcome, how it occurred, and why it is important. To ensure that details are forthcoming, evaluators prompt participants with questions such as, What happened? Why do you think this is a significant change? and What difference has it already made or will it make in the future? Evaluators also can follow up after they have read the stories to gather more details and verify their understanding. MSC stories have several advantages, including they allow "non-evaluation experts to participate, they are likely to be remembered as a complex whole, and they can help keep dialogue based on concrete outcomes rather than abstract indicators" (Dart and Davies 2003, 140).

Stories about the impact of interventions can infiltrate the collective memory of an organization and provide a base for dialogue about what is desirable in terms of expected and unexpected outcomes (Dart and Davies 2003). For example, environmental education organizations can seek feedback from funders about which stories are most in line with their funding goals and why. MSC stories also can uncover outcomes educators were not aware of and that are highly valued by participants, and which could be incorporated into subsequent programs and theories of change. Further, evaluators can use the outcomes described in stories to develop quantitative surveys, which provide information on how common and important various outcomes are for the larger group of program participants.

MSC is one of a number of participatory approaches to evaluation that engage participants and stakeholders in telling their stories and thus make them feel valued (Dart and Davies 2003). MSC also can be incorporated into program learning activities; for example, MSC questions could be assigned as the final reflective journal entry in an environmental education program (Zeegers and Clark 2014; see embedded assessment below).

Similar to MSC, appreciative inquiry is a participatory approach to evaluation with a focus on "positive organizational attributes that may fuel change" (Grant and Humphries 2006, 402). Appreciative inquiry entails using interviews to gather participant stories about what worked well in an environmental education program; evaluators engage in reflection and deliberation about the stories during and after the interview process (Grant and Humphries 2006; Reed 2007). In an urban agriculture intern program, a researcher conducted open-ended interviews with participants focused on how the program helped them build leadership, responsibility, and other youth assets (see chapter 14). After eliciting the interns' stories about what worked well in the program, the researcher shared the stories with the educators. Throughout the process the researcher expressed her appreciation of the program's ability to build youth assets, while also applying a critical lens through reflection, deliberation, and comparing her findings with those of previous researchers (Delia and Krasny 2018).

Embedded Assessment

Embedded assessment refers to program activities that are also used to assess program outcomes. Such activities could include concept mapping to learn about social-ecological systems, reflective journaling and media analysis to foster critical thinking, taking photos or creating drawings to foster sense of place, storytelling to foster intergenerational connections, and accounts of service learning to foster civic participation (Costantino and Greene 2003; Tal 2005; Heimlich 2010; Ardoin et al. 2014; Zeegers and Clark 2014). Generally the educator provides prompts to focus participants' attention on information relevant to program outcomes, such as asking participants to take photos of places that interest or are important to them, or write a journal entry about aspects of the program they want to remember five years from now. To analyze the participants' photos, stories, and other artifacts, evaluators code them for potential outcomes, like nature connectedness or place meanings, as well as any new outcomes or relations between activities and outcomes they did not anticipate. Often evaluators follow up with interviews to gain a more in-depth understanding of how a program has impacted participants. In one embedded assessment at an environmental education camp, participants were asked to take photos and write photo captions, which captured affective aspects of their experience, whereas in a journal writing assignment participants described program activities (Ardoin et al. 2014).

In reality, many evaluations integrate multiple methods. For example, you may use Likert scale questions on surveys to generate numbers of participants achieving levels of intermediate outcomes, and open-ended questions during interviews to discover quotes that enable deeper insight into the program. Mixed methods combine quantitative and qualitative techniques and thus use the strength of both to answer a broad range of questions (Johnson and Onwuegbuzie 2004). Regardless of the approach, I encourage environmental educators to contribute to and be involved in evaluations, both because you have valuable insights that an outside program evaluator lacks, and because you will be the most important user of results. Collaboration will provide learning opportunities for educators, evaluators, and funders.

Evaluation Steps

Not every environmental educator has training in evaluation and research. However, all environmental educators are intimately familiar with their programs and have the ability to observe and reflect on their program activities and theories of change. In other words, you are an expert, and you have the ability to think critically. So while sometimes it makes sense to partner with a professional evaluation firm and an organization has the resources to do so, if time and resources preclude this possibility, gaining insights into a program's effectiveness and areas needing to be changed is still possible. Here are general guidelines for conducting and learning from an evaluation—all of which can be part of a discussion with colleagues and other stakeholders. For specific details on evaluation steps, refer to the resources listed at the end of this chapter.

1. Clarify the *purpose* of your evaluation. How much emphasis do you want to place on demonstrating outcomes, discovering what things are working and what needs to be changed, professional development, and/or building a relationship with funders and other stakeholders?
2. Clarify your *theory of change*. Make sure to refer back to your intended outcomes and pathways for getting there, keeping in mind that evaluating your program and defining your theory of change can be an iterative process (see chapter 1). For example, if your intended outcomes are to change attitudes among adult participants, you might want to revise your theory of change and your evaluation given that adults' attitudes are difficult to alter, especially through a short-term program (see chapter 7). However, you may be able to increase systems- or action-related knowledge among adults (see chapter 6) and influence attitudes of young children—for example, attitudes toward specific behaviors or toward wildlife encountered during the program. Focus your evaluation on achievable outcomes and likely pathways to achieve them. You also can make adjustments in your activities and intended outcomes based on research outlined in the chapters of this book. In so doing, you have already engaged in the reflection and critical thinking that are key to every evaluation.
3. Decide *what* to evaluate. Likely your program does a lot of things. You may engage youth in community gardening or other civic ecology

practices in their neighborhood, and have multiple intermediate and behavior outcomes in your theory of change—for example, systems learning, nature connectedness, place identity, and intention to volunteer over the next six months. Do you have the capacity to assess all these outcomes and perhaps others that emerge during the evaluation, or should you focus on one or two? You can use the chapters of this book to help decide what outcomes to evaluate. You can focus on behaviors or collective actions, and/or on intermediate outcomes that are associated with behaviors and actions such as efficacy and environmental identity.

4. Decide on your evaluation *approach*. Although we often think about pre-/post-surveys and other more quantitative, objective means of evaluation, you don't need to limit yourself to these methods. In fact, sometimes these methods may steer you to misleading results, as when you show a short-term attitude change immediately after a program but research and theory would suggest those attitudes will not be sustained in the long term (Heimlich 2010). To truly assess changes using quantitative surveys, you will likely need to partner with an experienced evaluator or researcher, but other approaches can also provide insights into your program outcomes and need for adapting your program or theory of change.
5. Choose your evaluation *tools*. After you define your evaluation questions and approach, you will need to identify tools to measure the outcomes (Ernst et al. 2009). We have included sample evaluation survey questions and other tools in the outcomes chapters and appendix, which provide indicators of your outcomes. For example, indicators for well-being could be physical, such as ability to walk a certain distance, or cognitive, such as ability to find your way in a new neighborhood or natural area (see chapter 15). We also suggest qualitative tools to assess outcomes, including interviews that accompany surveys, and Most Significant Change and embedded assessment (see above). Feel free to adapt existing surveys and other tools for your own evaluations while acknowledging their source.
6. *Conduct* the evaluation and *analyze and reflect on* results. After you collect evaluation data, you will need to analyze it. To summarize quantitative data collected in surveys, you can use online survey tools such as SurveyMonkey and spreadsheet programs like Excel to compute averages (means) and other summary statistics. You can also use SurveyMonkey and spreadsheet programs to help visualize your data by making bar and line graphs and pie charts. To summarize findings from interviews, drawings, stories, and similar qualitative data, you can specify codes based on your desired outcomes and look for these codes in your data

(e.g., connectedness to nature). You may also want to take note of unexpected or "emergent" themes that appear in your data.

7. Engage in *discussion* and *communicate* about the results and their implications for your program and organization. Discussions with colleagues, funders, parents, and even program participants are an important part of learning through evaluation. You can also summarize results in a report, in a blog, on social media, or on a website.
8. *Adapt* your program based on discussion of and reflection on results. This may include revising your theory of change, changing program goals, adapting activities, and even considering new audiences, settings, and partners.

Ethics

Evaluation can benefit participants but also can pose risks to participants if you are not careful about the information you collect and how you collect and store it. Thus, it is essential that you follow ethical guidelines for working with human subjects; these guidelines are more stringent when working with children (Simmons 2004; UNICEF 2015). First, make sure participants freely participate in the evaluation rather than making it mandatory. Signed consent forms are necessary if you ask personal or sensitive information, use the results for publications, or use participants' names or photos; and for participants under age eighteen, parental consent is also required. Even if participants agree to take part in an evaluation, they have the right to refuse to answer any questions or withdraw at any time during the evaluation process. Second, when collecting information about individuals, keep the information confidential. You should remove names from surveys or interviews, store information in a secure location, avoid discussing information about specific individuals with others, and securely dispose of data when they are no longer needed. Finally, ensure participants' safety throughout the data collection process by not embarrassing or harming participants.

Challenges and Opportunities

> **Although I would agree that contemporary demands for accountability create some spaces for evaluation activities, I worry that accountability pressures can distort evaluation agendas and siphon valuable resources away from local, improvement-oriented evaluation activities.**
>
> (Greene 2010, 198)

Evaluators of environmental education programs face a number of challenges. The field of environmental education is broad; programs are conducted in settings ranging from schools to zoos to community organizations; and outcomes vary from knowledge to identity, sense of place, civic participation, and health and wellness, as well as direct impact on the environment. This makes it challenging to develop a one-size-fits-all approach to evaluation or to conduct large-scale studies across multiple sites. Programs also may lack clear objectives and a well-articulated, research-based theory of change, resulting in a mismatch between activities and program outcomes. Further, programs occur over a limited period of time, but the desired outcomes are often long-term changes in behavior, which are more likely to be the result of repeated or sustained interventions. Yet measuring long-term impact is fraught with the issue that other influences—social media, online news sites, church pastors, and family and friends—may also be impacting changes in behaviors. Finally, environmental education organizations face limited resources coupled with pressure to demonstrate positive impacts to funders and other stakeholders (Carleton-Hug and Hug 2010).

At the same time, a vibrant and growing environmental education community is determined to make a difference, and is increasingly networked globally (Civic Ecology Lab 2019b). Through educators partnering with trained researchers to conduct rigorous evaluations, and through educators continuing to make observations of their programs, reflecting on their practice and theory of change, being on the lookout for ways to improve their programs, and discussing what they learn with colleagues locally and globally, we can further a learning-through-evaluation culture in environmental education.

Evaluation Resources

Below are resources that may be helpful in planning and conducting evaluations.

My Environmental Education Resource Assistant (MEERA) is a website that offers step-by-step instructions and resources online for environmental educators (http://meera.snre.umich.edu/) (Zint, n.d.).

Evaluating Your Environmental Education Programs: A Workbook for Practitioners provides evaluation instructions, case studies, and exercises to help you develop your own evaluation (https://naaee.org/eepro/publication/evaluating-your-environmental-education-programs-workbook-practitioners) (Ernst et al. 2009).

Designing Evaluation for Education Projects provides information on evaluation methods and background (http://www.birds.cornell.edu/

citscitoolkit/toolkit/steps/effects/resource-folder/Designing%20Evalua tion%20for%20Edu%20Projects.pdf) (Simmons 2004).

Guidelines for Excellence in Environmental Education: The North American Association for Environmental Education (NAAEE) published a series of guidelines aimed at helping educators assess the quality of their environmental education programs based on research and on educators' experiences in the field. Although not evaluation per se, the guidelines can provide useful information for assessing how well your activities reflect the collective knowledge of researchers and practitioners (https://naaee.org/our-work/programs/guidelines-excellence) (NAAEE 1998–2016).

Part II

ENVIRONMENT AND BEHAVIOR/ACTION OUTCOMES

3

ENVIRONMENT, SUSTAINABILITY, AND CLIMATE CHANGE

Highlights

- Environmental education programs can have direct and indirect impacts on environmental quality, sustainability, and climate change.
- Direct impacts include changes in the biological and physical environment, such as increased biodiversity or water quality, or decreased levels of greenhouse gases in the atmosphere.
- Indirect impacts include environmental policies and management practices.
- Environmental education can integrate economic, social equity, and safety concerns related to sustainable development and environment-friendly climate adaptation.
- Educators can partner with scientists to assess changes in biodiversity using citizen science and other protocols, and can describe any new management practices or policies as a result of their environmental education programs.

Environmental quality, including sustainability and climate change mitigation, are often the ultimate outcomes of environmental education. Consideration of these outcomes shifts the focus from what participants think, feel, and do to the environmental impacts of their actions. Psychological and social factors like efficacy, identity, and social capital are intermediate outcomes or pathways to behaviors and collective actions that eventually lead to improved environmental quality. Environmental quality is important because of nature's intrinsic value and because a sound environment is necessary for humans to thrive. However, because of the impact humans already have had on the environment, for example on our climate, we are forced to adapt, preferably in ways that also mitigate future negative impacts. Thus, in addition to environmental quality, we discuss climate adaptation in this chapter. Resilience, another ultimate outcome that recognizes the need to adapt to change and incorporates social alongside environmental concerns, is discussed in chapter 16.

What Is Environmental Quality?

Environmental quality includes biological and physical, environmental management, and environmental policy outcomes, including those focused on climate change. When we broaden the environment to include humans as well as nonhuman nature, then sustainable development and social-ecological systems resilience, which integrate economic and social alongside environmental concerns, also can be considered part of environmental quality (Mebratu 1998; Walker and Salt 2006; Resilience Alliance 2009; see also chapter 16).

Physical and Biological Outcomes

Biological and physical outcomes include biodiversity, wildlife habitat, and water quality, as well as decreases in greenhouse gases, pollution, and waste. Biological and physical outcomes also include provision of ecosystem services, that is, services offered to humans by biodiversity and natural systems (Daily 1997; MEA 2005). For example, a rain garden provides the ecosystem service of filtering street water runoff and creates insect habitat.

Management and Policy

Change in government or NGO conservation or management practices can be an outcome of environmental education. Environmental action programs like Earth Force (n.d.) train participants to engage in the local policy process, and some citizen science programs result in changes in management or policy (Cooper et al. 2007). For example, citizen science data collected by volunteers spurred the British government to adopt a policy to increase farmland bird populations (Kobori et al. 2016).

Sustainable Development

As environmental protection was increasingly pitted against economic development in the late 1980s, the United Nations released *Our Common Future*—also known as the Brundtland Report (Brundtland Commission 1987)—which proposed a vision of sustainable development.

> Sustainable development is development that meets the needs of the present without compromising the ability of future generations to meet their own needs. It encompasses two key concepts:
>
> - The concept of "needs," in particular, the essential needs of the world's poor, to which overriding priority should be given; and
> - The idea of limitations imposed by the state of technology and social organization on the environment's ability to meet present and future needs. (Brundtland Commission 1987, 41)

The initial report was followed in 1992 by the action plan Agenda 21 for sustainable development, which emphasized the role of education and participation in bringing about a just and sustainable society (United Nations 1992). In September 2015, the United Nations adopted seventeen "sustainable development goals," with education being a goal in itself and playing a role in the implementation of the other goals (United Nations 2015).

In short, the notion of sustainable development proposes that society can meet goals for economic development and equity, while also sustaining natural systems. In this human-centered view, natural systems provide ecosystem services, such as food, fiber, and clean water, on which the economy and human society depend. Ecosystems are kept healthy in order to provide ecosystem services to humans and to prevent wastes from irreversibly imperiling the environment (Moldan et al. 2012), whereas the intrinsic value of nature is largely ignored.

Although early work in environmental education addressed social issues (Stapp et al. 1996), these issues assumed center stage alongside environmental issues with the advent of "Education for Sustainable Development" in the 1990s (UNESCO 2002; Wals 2012; UNESCO, n.d.). Education for Sustainable Development is defined as a "process of learning how to make decisions that consider the long-term future of the economy, ecology and social well-being of all communities" (UNESCO 2002, 10). Whereas Education for Sustainable Development has assumed prominence globally, it bears many similarities to environmental education and shares many of environmental education's strengths and shortcomings (Huckle 1993; Monroe 2012). Further, sustainable development outcomes are broad and generally measured on a national or regional scale (e.g., proportion of population using safe drinking water) (Moldan et al. 2012). Because of environmental education's longer history and its incorporation of the intrinsic value of nature, alongside the difficulties of measuring sustainable development outcomes of educational programs, I focus on environmental education in this book. However, many of the concepts can be applied to Education for Sustainable Development and related approaches.

Climate Change: Adaptation and Mitigation

In the era of climate change, a single focus on reducing greenhouse gases may fail to address immediate threats that imperil communities, like rising sea level, drought, and wildfire (Krasny and DuBois 2016; Krasny, Chang, et al. 2017). In particular, climate adaptation, or adjustments in ecological, social, or economic systems to address the impacts of climate change (UNFCCC 2014), suggests learning to live with rather than attempting to prevent climate change. In contrast, mitigation refers to efforts to reduce or prevent greenhouse gas emissions (United Nations Environment Programme, n.d.). Numerous United Nations and other educational programs are already focusing on climate adaptation—for example, teaching children to change their walking routes to school to avoid frequently flooded areas (UNICEF 2012).

The question arises: Is education that accepts climate change as a given and incorporates how to adapt to a climate heavily impacted by humans consistent with the environmental quality goals of environmental education? On the one hand, a focus on adaptation could divert attention from important consumer, transportation, and energy choices consistent with climate change mitigation. On the other hand, a focus on mitigation alone may not protect people from flooding and other dangers, some of which children may face daily while simply walking to school (UNICEF 2012; Krasny, Chang, et al. 2017).

Adaptation actions can include anything from building infrastructure such as seawalls or artificial oyster reefs to abate storm wave action, to individual behaviors such as home disaster preparedness, changes in farming practices, and new policies like regulating construction along shorelines (Matthews and Waterman 2010). Some adaptation actions work counter to environmental quality. For example, building concrete seawalls may protect against sea level rise at one location, but negatively impact natural processes of sediment deposition and reduce shellfish habitat and biodiversity. Alternatively, restoring natural barriers such as oyster reefs or dunes can foster biodiversity and provide pollution filtering, food, and recreational ecosystem services. Three general categories of adaptation options are physical/structural, social, and institutional. Within each of these categories, a subset of strategies improves environmental quality while protecting communities from climate-related disaster (IPCC 2014).

In the physical/structural category, *ecosystem-based* adaptation is the subset of adaptation strategies that enhance biodiversity and ecosystem services. An example of ecosystem-based adaptation is restoring and maintaining wetlands to buffer against sea level rise and storm impacts, to sequester carbon, and to provide fish habitat (IPCC 2014). In cities, green infrastructure, such as green roofs, porous pavement, and parks and community gardens, can alleviate climate-related flood risk and high temperatures, and has collateral benefits such as food production, recreation, and education.

Social adaptation options reduce risks among vulnerable populations and address social inequities. *Community-based adaptation* is the subset of social adaptation options that are locally driven and operate "on a learning-by-doing, bottom-up, empowerment paradigm that cuts across sectors and technological, social, and institutional processes" (IPCC 2014, 847). Civic ecology and educational practices that entail collaborations among community-based organizations and government to restore streams, plant trees, and install bioswale gardens are examples of community-based adaptation strategies that reduce flooding and heat wave risk while enhancing environmental quality (Krasny, Lundholm, et al. 2013; Krasny and Tidball 2015). Such practices can foster interactions among volunteers and scientists and thus incorporate diverse knowledge, experiences, and perspectives (IPCC 2014).

Institutional adaptation strategies entail changing policies. For example, government might provide financial incentives for people to move away from flood zones or areas subject to forest fires. Government and NGO stakeholders might then protect such areas to provide flood mitigation, wildlife habitat, and other ecosystem services (IPCC 2014).

In short, ecosystem-based, community-based (including education), and some institutional adaptation strategies are consistent with environmental quality and climate mitigation goals.

Why Is Environmental Quality Important?

Environmental quality is important because nature has intrinsic value and because a healthy environment is critical to human health and well-being. The earth's biodiversity, including humans, faces an existential threat brought about by human-caused greenhouse gas emissions leading to climate change. These threats include drought, forest fires, heat waves, flooding, and sea level rise. Other threats to humans and nonhuman life include habitat loss, plastics pollution, and chemical contamination. Addressing these problems is the ultimate goal of most environmental education programs.

How Does Environmental Education Enhance Environmental Quality?

Civic ecology, citizen science, place-based, energy-use feedback, and environmental action approaches often focus more directly than other environmental education programs on enhancing environmental quality. They do so by impacting biodiversity and habitat; energy, water, and waste; management and policy; and environmentally friendly adaptation.

Biodiversity and Habitat

Programs in which participants work directly to increase numbers and types of plants, insects, birds and other animals, or to enhance wildlife habitat, can increase biodiversity. For example, in a primary school in suburban Stockholm, students rescued nearly five thousand salamander newts that had fallen into a concrete pool by transporting them to a nearby pond, allowing the local salamander population to remain stable over a ten-year period (Barthel et al. 2018). Citizen science programs that include conservation actions, such as Habitat Network, in which homeowners add nectar plants and bird feeders to their yards, can enhance wildlife populations. Habitat Network participants form local or virtual groups that adopt common goals and through their collective efforts create a patchwork of bird habitats (Dickinson et al. 2013). In Ota City, Japan, citizen science volunteers observed that little terns (*Sternula albifrons*) preferred to nest on sites with a white background, and, after confirming this observation through conducting an experiment in collaboration with scientists, placed fragments of white shells on roofs (Kobori et al. 2016). School and community gardening, tree planting, and other civic ecology education programs in which youth engage directly in restoration or gardening also have the potential to increase wildlife habitat, biodiversity, and related ecosystem services (Doyle and Krasny 2003; Andersson et al. 2007; Blair 2009; Krasny and Tidball 2009a; Krasny et al. 2009; Beilin and Hunter 2011; Krasny, Lundholm, et al. 2013).

Energy, Water, and Waste

Environmental education programs that directly reduce energy use often take place in school or dormitory settings, where one can readily measure energy consumption using building utility meters rather than rely on self-reported energy use. Additionally, changes in meter readings for energy and water use can be compared across dormitories within a college, and aggregated across dormitories to rate different universities. Use of digital feedback documenting energy use, including in competitions where students compare how well their dormitory or university is doing relative to their peers, has led to reductions in dormitory and K–12 school energy consumption (Petersen et al. 2015; Petersen et al. 2017). Competitions also have been used to reduce waste among university dormitories, which can be measured as increased rates of recycling or decrease in weight of waste sent to landfills (Short 2010; Petersen et al. 2015).

Policy and Management

Environmental action programs commonly engage youth in researching an issue and then advocating with local authorities for policy change (Jensen and Schnack 1997; Schusler et al. 2009; Johnson et al. 2012). Although often these programs stop at presenting information to officials and do not have immediate policy impacts, in some cases actual policy change ensues. For example, in a Chinese university, student environmental club members identified an opportunity to reduce use of carryout containers, and persuaded the cafeteria staff to curtail plastics use (Yu 2018a). In Brooklyn, New York, youths' discovery of a high number of lead poisoning cases near the Williamsburg Bridge was one factor in the city temporarily stopping sandblasting until the lead could be contained. And in rural New York, students identified a highway department salt pile as a source of streamwater contamination, and their subsequent recommendations helped spur a town resolution to construct a salt storage shed, thus reducing salt runoff into the stream (Mordock and Krasny 2001). Similarly, students in New Hampshire monitored car idling in the school drop-off area, which led to the school adopting a no-idling policy, thus reducing car emissions; and water-quality monitoring programs have led cities to repair sewer systems, ban the use of lawn chemicals on school property, and pass local water-quality ordinances (Johnson et al. 2012).

The popular citizen science program eBird has generated massive reams of data that have been used in conservation planning, policy, species management, and habitat protection. Examples include using bird population data in developing the IUCN Red List of Chile's threatened bird species, placing nest boxes to maximize waterfowl populations in Canada, developing plans to protect sites of high conservation value in New Zealand, locating communications infrastructure to minimize impacts on birds, and listing a subspecies as threatened in the United States (Sullivan et al. 2017). In US and Australian cities, stewardship volunteers collecting data appear to have influenced local housing or broader municipal policy (Beilin and Hunter 2011; Silva and Ramirez 2018). These projects are evidence of how environmental education programs in which participants collect data can become part of larger efforts engaging volunteers, NGOs, and government agencies that result in significant environmental impacts.

Environmentally Friendly Adaptation

Multiple pathways exist for environmental education programs to enhance the ability of communities and local ecosystems to adapt and transform, including in response to climate change. Often this entails joining forces with larger efforts, such as New York City's Rebuild by Design initiative, where youth may participate in planning and helping to construct green infrastructure such as raised areas along waterfronts planted with salt-tolerant trees and shrubs (Rebuild by Design 2015). Similarly, after Hurricane Sandy struck New York City, youth participated in oyster restoration projects designed to protect shorelines from future flooding, in partnership with nonprofit organizations, universities, and state government (DuBois and Krasny 2016). Youth might also work with local leaders to develop a city resilience strategy (100 Resilient Cities 2018) or a plan to obtain certification as a Climate Smart Community (DEC, n.d.).

Assessing Environmental Outcomes

Assessing environmental outcomes of environmental education programs is challenging for several reasons, perhaps most importantly because environmental education is only one factor among many that lead to changes in biodiversity, greenhouse gas emissions, or environmental management and policy. One might think of environmental education as one tool in the toolbox of improving environmental quality, or perhaps more accurately, given the importance of partnerships in effecting change, one node in a network of environmental organizations (Svendsen and Campbell 2008). Adding to the difficulty of measuring environmental impacts is the fact that environmental changes often occur over years and also may be impacted by unexpected events (e.g., floods, fires). Further, measures of actual changes, such as in water quality, vary over time and space, and we may lack comparison data from before a program (Short 2010). One approach is using proxy measures of environmental quality, such as reduction in energy consumption, rather than direct measures of greenhouse gases (Johnson et al. 2012).

The Environmental Education Performance Indicator was developed to estimate different types of environmental impacts even in the absence of pre-program comparative data. This indicator estimates the effectiveness of an environmental education intervention by comparing the actual environmental impact resulting from student actions to an estimate of what might have occurred without the program and what might be an optimal outcome under ideal conditions. It has been used to estimate outcomes ranging from tons of toxic airborne emissions not being released as a result of high school students blocking the construction of a waste incineration facility to changes in policy and biodiversity (Short 2007).

Biophysical Outcomes

Students have monitored indoor air quality such as particulate levels and mold (Johnson et al. 2012) and used citizen science protocols to measure water quality (Shirk et al. 2012) and populations of dragonflies and other organisms (Primack et al. 2000; Kobori et al. 2016). Youth from the community organization Rocking the Boat in New York City worked with scientists to grow mussels and seaweed on rafts designed to filter nitrates released from a wastewater treatment plant, and monitored seaweed growth, mussel nitrate levels, and water quality as indicators of the rafts' ability to filter pollution. They also partnered with the Department of Parks & Recreation and nonprofit NY/NJ Baykeeper to monitor the growth and mortality of oysters in cages they installed in the Bronx River, as indicators of biodiversity and a proxy measure of pollution filtration (RTB 2012). Other measures include area in desired land use (e.g., bird habitat, restored wetland, or vegetable gardens, which could be measured using maps or Google Earth); presence of management practices (e.g., composting, energy consumption); or measures of outputs such as pounds of vegetables produced (Krasny, Russ, et al. 2013). Tree planting ecosystem services can be measured using i-Tree software (US Forest Service, n.d.) and its simpler adaptation, the online National Tree Benefit Calculator (n.d.).

Management, Policy, and Adaptation

Changes in management or policy as a result of participants' actions can be used as proxy measures to assess environmental outcomes. For example, helping to persuade a government agency to change its plans for development of critical bird habitat, a school to implement more meat-free meals, or a city to reintroduce alewife fish to the Bronx River (Goncalves 2009; Kudryavtsev 2013) can be readily documented and could have far-reaching impacts on environmental quality.

4

ENVIRONMENTAL BEHAVIORS

It is not enough for environmental education to promote action for the environment: It needs to emphasize the most strategic actions. . . . The effect of private actions is limited unless it is combined with organizing for collective public change. If environmental educators confine themselves to fostering private sphere environmentalism, they may in fact be leading students astray.

(Chawla and Cushing 2007, 438)

Highlights

- Environmental behaviors are "any actions taken by an individual or a group that benefit the natural environment, enhance environmental quality, or promote the sustainable use of natural resources" (Larson et al. 2018, 871–872).
- Lifestyle, stewardship, citizenship, and social environmental are categories of environmental behaviors.
- Spillover occurs when engagement in one environmental behavior encourages engagement in another environmental behavior.
- Eating less meat and dairy, reducing food waste and car and airplane travel, and turning down the heat or air conditioning are among the most effective lifestyle behaviors in reducing greenhouse gas emissions.

Environmental behaviors are "any actions taken by an individual or a group that benefit the natural environment, enhance environmental quality, or promote the sustainable use of natural resources" (Larson et al. 2018, 871–872). All kinds of behaviors fit into this definition. Typically you might think of recycling, turning down the heat, or turning off the lights. But how about donning an electric jacket or heating pad that warms your body without having to heat the whole house? Or plalking—picking up litter while walking? And many might not realize that eating less meat, having fewer children, and reducing food waste and air travel are more effective in reducing greenhouse gases than are commonly recommended energy saving behaviors like turning off lights (Wynes and Nicholas 2017; Lacroix 2018; Drawdown, n.d.). Advocating for a climate policy, volunteering to collect data on seabirds, or becoming active in a friends of parks group are public sphere or citizenship behaviors, also referred to as collective actions (see chapter 5). In short, environmental educators can focus on any number of environmental behaviors, all with different impacts on the environment (Gatersleben 2013; Larson et al. 2015).

The good news is that the wealth of environmental behaviors means everyone can find a place to pitch in. The challenge is finding ways environmental education can influence such diverse behaviors, with a focus on those behaviors that most effectively reduce greenhouse gases, plastics use, and other sources of contamination, and preserve biodiversity. We can also consider stewardship behaviors where humans positively impact local environments and communities. Importantly, we can consider actions to influence policies, which in turn impact millions of individual behaviors. Whereas in part 3 of this book we talk about pathways to influence environmental behaviors, here we cover types of environmental behaviors, their interactions, and their effectiveness in addressing environmental problems.

Making Sense of the Wealth of Environmental Behaviors

Given the wealth of environmental behaviors, it is not surprising that there are multiple ways to categorize them (table 4.1). Perhaps the most common classification is separating behaviors related to individual lifestyles and those intended to influence policy. Lifestyle behaviors are often the focus of environmental education. They are things we do every day, such as turn down the heat and recycle, and the individual choices we make about transportation, food, energy, waste, water, and purchases (Stern 2000b; Gifford 2014; Kurisu 2015; Larson et al. 2015; Truelove and Gillis 2018). Lifestyle behaviors take place in homes, offices, schools, and public spaces, but factors influencing these behaviors differ depending on where they take place. For example, at home our actions are often guided by what parents and other family members think we should do, whereas what we see others doing has a major influence when we are in public (Cialdini 2007; Kurisu 2015; see also chapter 9). Lifestyle behaviors are important because if enough people adopt them—especially those behaviors that are most effective at reducing greenhouse gases or at achieving other environmental outcomes—we can collectively make a difference for the environment (Dietz et al. 2009).

TABLE 4.1 Environmental behaviors

BEHAVIOR	DEFINITION	EXAMPLES
Lifestyle (also called private sphere behaviors)	Daily behaviors and consumer choices (Stern 2000; Larson et al. 2015)	Individual choices about transportation, food, energy, waste, and purchases, such as using personal heating devices (e.g., small electric rugs or electric jackets) and lowering heat, reusing consumer goods, recycling, reducing car/plane trips, and replacing meat/dairy with vegetables, nuts, and grains
Citizenship (also called public sphere or activist behaviors)	Behaviors intended to influence policy makers and change policies (Stern 2000; Larson et al. 2015)	Voting, petition signing, letter writing, protesting, donating money
Stewardship (also called non-activist public sphere behaviors)	Hands-on activities taken to directly improve habitat, increase biodiversity, and enhance ecosystem services (Larson et al. 2015)	Civic ecology practices including community litter cleanups, community gardening, tree planting, invasive species removal, and oyster or mangrove restoration (Krasny and Tidball 2015); backyard habitat enhancement; citizen science biodiversity monitoring (Monarch Watch 2014)
Social environmental	Interaction or communication to inform or teach others about conservation and environmental behaviors (Larson et al. 2015)	Taking groups on educational nature hikes, posting environmental messages on social media
Climate change mitigation	Individual or group behavior that reduces greenhouse gas emissions (IPCC Core Writing Team 2014)	Adopting renewable energy, including installing solar panels on individual homes, choosing community solar, and supporting wind energy projects
Ecosystem-based adaptation	Individual or group behavior that helps a community adapt to changes brought about by climate change in an environmentally friendly manner (IPCC 2014)	Restoring dunes and oyster reefs to protect shorelines, installing solar-powered charging stations in public spaces to enable cell communications when power is out

(cont.)

BEHAVIOR	DEFINITION	EXAMPLES
Positive spillover	Environmental behavior influenced by having engaged in previous environmental behavior (Nilsson et al. 2017)	Participating in a litter cleanup leads to participating in subsequent cleanups, reusing household objects, and advocating for plastic straw ban
Negative spillover (also called rebound behavior)	Non-environmental behavior influenced by having engaged in previous environmental behavior (Nilsson et al. 2017)	Justifying taking an airplane trip by having reduced car use; spending money saved by reducing energy on consumer goods

Changing individual lifestyle behaviors, however, is not enough to solve environmental problems. Policies that both reward positive behaviors and sanction polluting behaviors are also critical. When humans attempt to influence policy, for example through advocacy, voting, and civil disobedience, they engage in environmental citizenship behaviors (see also chapter 5). Any one new regulation or law can have a massive impact on lifestyle behaviors (e.g., passing a fee on single-use plastic bags) or the behavior of industrial and other polluters (e.g., regulations on carbon emissions from vehicles or power plants) (Stern 2000b; Larson et al. 2015). Action competence and similar action-oriented environmental education programs are designed to build participants' ability to influence policy, for example at their school or local municipality. Participants in such programs research an issue, come to a consensus on a problem they wish to address, and then communicate with school administrators or elected officials about the need to change the related policy (Jensen and Schnack 1997; Volk and Cheak 2003; Chawla and Cushing 2007; Schusler and Krasny 2010; Schusler 2014; Earth Force, n.d.). In countries where citizens have limited ability to vote or otherwise influence government policy, they still may be able to influence school, university, or workplace policies, for example by advocating for reusable dishes in the cafeteria. Further, some lifestyle or "private sphere" behaviors that have large environmental impacts are generally considered outside the realm of government regulation—for example, veganism and small family size. Social media (Ballew et al. 2015) and NGOs may play a role in changing these behaviors.

Whereas lifestyle behaviors involve reducing humans' negative impacts, stewardship focuses on humans as a positive force acting to improve degraded environments. Examples include people converting trash-strewn vacant lots to productive gardens, and habitat improvement on public and private lands (Larson et al. 2015). These are called "win-win" behaviors because they have positive impacts on human well-being and the environment (Kurisu 2015).

Social environmentalism encompasses talking to others about environmental issues and being active in environmental groups (Larson et al. 2015). An example is young people talking with family members about reducing food waste or getting involved with a local group that distributes discarded food from restaurants to homeless people. Given the potential for social media to influence environmental behaviors, young people might join a Facebook group or use an app that shares resources on converting plastic bottles to useful household goods, reports illegal dumping, or otherwise promotes environmental behaviors (Ballew et al. 2015; Wamuyu 2018). Importantly, we can also use our own social networks to influence others and thus expand the impact of our individual behaviors (Krasny 2019).

Climate change behaviors can be categorized as mitigating human impacts on the climate by reducing greenhouse gases and adapting to more severe storms, droughts, heat waves, and other realities of climate change (IPCC Core Writing Team 2014). It is important to identify environmentally friendly means of adapting to climate change, as some adaptations, such as erecting concrete barriers to prevent flooding or using more air conditioning, can have negative environmental impacts. Ecosystem-based adaptations mitigate greenhouse gases while helping communities adapt (see chapter 3). An example is installing bioswale gardens whose plants take up CO_2 while absorbing water and thus reduce flooding during heavy downpours. Urban community gardens provide a source of local fresh food, which reduces transport costs and encourages vegetable rather than meat consumption, while also providing green and shady spaces to get away from the heat (IPCC 2014; Krasny and DuBois 2016). When people act as community stewards to plant and maintain such gardens, they build connections and stronger communities better able to adapt to future change in an environmentally friendly manner (Manzo and Perkins 2006). Social-ecological resilience comprises adaptation as well as devising transformative, environmentally friendly innovations to address climate and other types of change (Folke et al. 2002; Walker and Salt 2006; Resilience Alliance 2009; see also chapter 16).

Effectiveness of Environmental Behaviors

> **Most individuals misjudge the relative importance of pro-environmental behaviors; individuals underestimate the climate impact of meat eating and overestimate the impact of excessive packaging, littering, and turning off lights.**
>
> (Lacroix 2018)

"Why do I want to change students' behaviors?" is perhaps the first question that should be asked in embarking on an environmental education program. An obvious reason is to improve the environment, but that's too broad to be useful. Perhaps the biggest threat facing humanity is climate change, so environmental education programs seeking to improve the environment might consider actions to reduce greenhouse gases. But there are plenty of other reasons to choose from—enhancing participant health and well-being, reducing waste, or stewarding urban open space, to name a few.

Once a desired environmental outcome has been identified, knowing which behaviors are most effective in achieving that outcome is crucial (Fremerey and Bogner 2014; Drawdown, n.d.). The effectiveness of behaviors depends on multiple factors, such as how local energy is produced or how open government officials are to citizen input. Further, effectiveness can vary depending on one's environmental goal; for example, banning single-use plastic bags helps reduce plastic litter, but producing sturdy reusable bags requires more resources and thus may increase resource consumption if the stronger bags are used only a few times. The effectiveness of behaviors also changes over time as a result of humans adopting new technologies. For example, once you install LED lights, lighting consumes relatively little energy, so you might direct participants' attention to other energy-saving behaviors. And if your community has bought into solar energy, then transportation and food choices become priorities for reducing greenhouse gas emissions (Drawdown, n.d.). Additional factors, such as your audience, the feasibility of changing a particular behavior, and organizational support, go into your final decision about which behaviors to target (Steg and Vlek 2009).

Of all the possible environmental outcomes, there seems to be the most information on the effectiveness of different actions to reduce greenhouse gas emissions. Below we also offer guidelines on plastic waste reduction, given the enormity of the global plastics problem.

Reducing Greenhouse Gases

> **The livestock supply chain emits 44 percent of the globe's human caused methane, according to the U.N.'s Food and Agriculture Organization—and a large slice of that comes from cattle's methane burps. So anything you could do to cut down on cow belching would, literally, help save the planet.**
>
> (Mooney 2015)

Before writing this book I had no idea that the lifestyle behavior deemed to have the largest impact on greenhouse gases is meat and dairy consumption. Shifting to a vegetarian diet is ten times more efficient in reducing greenhouse gases than reducing room temperature by 1°C. Reducing the number of meat days from seven to six per week (one vegetarian day per week) is slightly more effective than reducing room temperature by 2°C. Reducing food waste is also ranked among the top ten climate change solutions; nearly one-third of food produced is wasted, ending up costing not only billions of dollars but gigatons of greenhouse gases (Faber et al. 2012; Lacroix 2018; Drawdown, n.d.). Recently, the Canadian government incorporated environmental concerns into its healthy food guidelines, stating, "In general, diets higher in plant-based foods and lower in animal-based foods are associated with a lesser environmental impact. . . . Planning meals and food purchases can also help decrease household food waste" (Canada 2017).

After reducing meat and dairy consumption and food waste, changing transportation behaviors is the next most effective category of behaviors an individual can take to reduce greenhouse gases. This includes daily behaviors like walking, biking, scootering, or using public transport instead of driving; not speeding up or slowing down quickly if you have to drive a car; and major purchasing decisions such as buying an electric, hybrid, or small car. Other transportation-related behaviors effective in reducing greenhouse gases include holding virtual meetings, telecommuting, and reducing miles driven by delivery trucks through grouping online purchases and avoiding one-day delivery (which requires more and longer trips for trucks). For higher-income households with more disposable income, transportation emissions can exceed those from food, owing largely to more frequent air travel, which is a major carbon emitter (Faber et al. 2012; Denchak 2017; Lacroix 2018). Thus, reducing transportation emissions is particularly effective in reducing carbon footprint in higher-income households. If this is not possible, buying carbon offsets such as those offered by airlines and nonprofit organizations is an option, providing the programs are transparent, verifiable, and effective (Palmer 2016).

The third category of behaviors relates to building temperature. It entails reducing room temperature during winter and turning up the thermostat in air-conditioned buildings in summer. Small space heaters are energy hogs. Alternatives that consume less energy and provide the comfort of higher room temperatures include electric foot warmers and other personal heating and cooling devices (Faber et al. 2012; Veselý and Zeiler 2014; Zhang et al. 2015; Veselý et al. 2017; Lacroix 2018).

A general guideline for effectiveness of personal home energy consumption is "heat, light, electronics." Thus, after addressing heating and cooling, changing to LED lightbulbs is the next most important thing we can do; LED lightbulbs reduce energy for lighting by up to 80 percent over older incandescent bulbs. Using smart power strips, which shut off power to electronics when they are not in use, can save up to 75 percent of the energy consumed by household electronics (EnergySave 2018).

Other impactful behaviors include proper disposal and replacement of older cooling units, such as air conditioners and refrigerators. HFC, a common coolant used in refrigerators and air conditioners, can be up to thirteen hundred times more potent than CO_2 in raising atmospheric temperatures. (Scientists estimate that clamping down on HFC use could avert 0.5°C of future warming; Reese 2018.) Although this and other impactful solutions, like installing solar panels, may be out of reach for many environmental education participants, they could be the focus of environmental action programs—or citizenship behaviors—that seek to influence policy such as tax breaks or incentives for renewable energy. Additionally, becoming familiar with less obvious climate solutions—like creating educational opportunities for girls in developing countries and family planning—may stimulate discussion of the inextricable links between environmental and social systems and issues of equity and justice (Drawdown, n.d.) and thus provide an opportunity to foster systems thinking and critical ecological literacy (see chapter 6).

Author Reflections

One winter day as I was sitting in my house, the thought dawned on me: Why am I heating this whole house when really the chairs, tables, and even the water pipes would be quite happy with 50°F (10°C) or even colder temperatures? My cold-blooded daughter was about to come home for the holidays, and I knew she would crank up the wood stove to sauna-like temperatures. That's when I hit on the idea of personal heating devices, and lo and behold, I searched online and found jackets with electric heating elements. I purchased one for my daughter for Christmas. She was quite happily snug as a bug in the jacket and refrained from heating up the whole house to her comfort level. I then discovered that placing a rubber hot-water bag used for pain relief behind the small of my back allows me to be very comfortable in a cool house. Personal electric rugs that warm feet also allow me to remain comfortable with lower building temperatures (cf. Zhang et al. 2015).

Reducing Plastics

Of course climate change is not the only environmental risk. Plastic wastes have been accumulating at an unprecedented rate, interfering with navigation and fishing in rivers, turning once pristine beaches into unsightly dumps, killing marine life that consume or get caught in plastics, and breaking down into small fragments called microplastics that could pose as yet unknown threats to humans and ocean life. Although many environmental education programs focus on recycling, reducing plastics use—in particular of onetime-use plastic items—is more effective in addressing the plastics problem. Educators and program participants can avoid bottled water when clean water sources are available, refuse to use plastic straws and plastic cutlery, bring their own coffee cup to Starbucks or Dunkin' Donuts, tuck reusable containers into their packs or purses for takeaway from restaurants, avoid hygiene products containing microbeads, buy in bulk or in larger containers, and avoid heavily packaged foods (Engler 2016).

Students studying plastics or energy consumption should know that not buying or using an item is the most effective means of reducing resource use, reusing items (multiple times) is generally the second most effective, whereas recycling requires greater expenditure of energy and raw materials. The mantra "reduce, reuse, recycle" captures ways to reduce plastics in order of effectiveness.

Author Reflections

I facilitate the Global Environmental Education Facebook group (Civic Ecology Lab 2019b), a platform where people share sustainability practices from around the world. I have noticed a lot of recent posts about the global plastics problem, including innovations to address this issue. It seems that plastic bottles have become a "natural resource" in some countries. They are used for making new products, like hanging planters, laundry baskets, and even ecobricks used in furniture and buildings.

Modeling Effective Behaviors

Note that environmental education programs can model effective behaviors. Given the impact of eating meat and dairy, environmental education programs could replace these items with other protein sources. They can also demonstrate green technologies like solar panels, compost systems, and water refill stations (see choice architecture, chapter 12). Further, environmental educators can include "effectiveness knowledge" in their programs. For participants who already want to do something to help the environment, knowing which actions are most effective helps prioritize behaviors.

Regardless of which behaviors you adopt personally or include in your environmental education programs, it is important to keep up-to-date on your "effectiveness" knowledge as technologies and agricultural practices change. As new refrigerators come on the market, HFCs are being replaced with more environmentally friendly coolants. And research is being conducted on using seaweed to reduce belching by cows—cow burps being a major global source of the potent greenhouse gas methane (Mooney 2015). Should cow burping decline, it could change the calculus on eating beef and dairy relative to other behaviors like reducing car use.

Spillover Behaviors

You might wonder: Will a participant who has adopted an environmental behavior during my program engage in additional environmental behaviors? Spillover refers to the likelihood that engaging in one environmental behavior influences engaging in another environmental behavior (Nilsson et al. 2017). A behavior can either increase or decrease the probability of repeating the same behavior at a later time or in a different setting, or of conducting a new behavior. For example, citizen scientists may start by participating in a one-day seabird monitoring activity, and then volunteer to monitor the same stretch of beach monthly. They may join in a second data collection effort at another site or focused on other species. Their data collection efforts may spill over into stewardship behaviors like protecting shorebird habitat by installing fences around nesting areas or cleaning up debris. These behaviors may also spill over into lifestyle behaviors like reducing onetime plastics use as well as social environmentalism behaviors like sharing their monitoring findings and stewardship experience with friends. Finally, the initial data collection and other behaviors may spill over into citizenship behaviors where participants use their findings to advocate for conservation policies (Larson et al. 2015; Haywood et al. 2016; Nilsson et al. 2017).

When working with people who do not see themselves as environmentalists, environmental educators can attempt to change individuals' self-perceptions to influence spillover behaviors. According to self-perception theory (Bem 1972), people who perform an environmental behavior may not view that behavior as environmental, either because they are doing it for a nonenvironmental reason (e.g., turning down heat to save money) or because the behavior is so common (e.g., not littering). An environmental educator can "cue" such behaviors as environmental—that is, point out that the behavior just performed is an environmental behavior. This in turn helps people perceive themselves as environmentally conscious and increases the likelihood they will perform other environmental behaviors (Cornelissen et al. 2008).

In addition to cueing, environmental educators can use foot-in-the-door strategies to foster spillover behaviors. If your program participants perform a small environmental action, like remember to put their food wastes in the compost bin, you can describe their behavior in terms of its environmental importance and provide feedback that helps invoke participants' environmental identity. Participants will then be more likely to repeat the behavior or to conduct additional behaviors consistent with their nascent environmental identity (Nilsson et al. 2017; see also chapter 11).

Both cueing and foot-in-the-door strategies suggest alternatives to focusing on attitudes and other cognitive and affective precursors of environmental behaviors, or on negative messaging about "bad" behaviors. Instead, educators recognize existing "good" behaviors, engage participants in simple enjoyable actions, and provide positive feedback that encourages additional environmental behaviors (Cornelissen et al. 2008; Haywood et al. 2016).

Negative spillover behaviors occur when individuals think they have already done something for the environment and thus do not need to take additional action. This thinking might go: I turn down the heat in my house and therefore I can take longer showers. Or the money I saved on purchasing solar power I will spend on a flight to the Caribbean. This is known as the rebound effect—in this case, adopting energy-saving technologies saves money that is spent on other energy-consuming activities (Nilsson et al. 2017; Lacroix 2018).

Spillover behaviors raise the question of feedback pathways in environmental education. Whereas we normally think that if we can change identity, attitudes, or efficacy we will influence behaviors, what happens when we start with a behavior, and that influences identity, attitudes, or efficacy? For example, you invite your cousins with no interest in the environment to join you in a beach cleanup, and they feel good about their experience and begin to think of themselves as environmentalists. Because this new environmental identity and other factors then go on to influence additional behaviors, we can think of this process as a feedback. A feedback occurs when engaging in environmental behaviors influences identity, efficacy, or another intermediate outcome, which in turn makes subsequent behaviors more likely.

Pathways to Influence Environmental Behaviors

Three general pathways are used to influence environmental behaviors: personal, social, and structural (Ballew et al. 2015). In this book, we focus largely on personal and social pathways. Note that simply engaging program participants in environmental behaviors—whether that might be composting at an environmental education center or participating in a volunteer stewardship project—is also a pathway to future behaviors.

Personal pathways leading to environmental behaviors include both cognitive, or thinking, and affective, or emotional, components. At one end of the spectrum is knowledge, a cognitive component. Additional cognitive factors include values, or general principles that guide our behavior (see chapter 7); self-efficacy, a belief about the change that one is capable of making (chapter 10); and norms, or beliefs about what others are doing or think we should do or what we think is moral (chapter 12). Other pathways, including attitudes, sense of place, and identity, are both cognitive and affective. For example, attitudes include beliefs and feelings about an object, such as nature, a particular park, or a stewardship behavior (chapter 7). Sense of place includes cognitive place meaning and affective place attachment (chapter 8). Finally, nature connectedness (chapter 8) and sense of hope (chapter 15) emphasize the affective—feelings about our relationship to nature and feelings of hopefulness.

Social pathways to environmental actions include social capital (chapter 13) and sense of community (Chavis and Wandersman 1990). Interestingly, many personal pathways also have social components. For example, people and communities can have a sense of place; people have self-efficacy and collective efficacy; people have individual and collective identities; and social and personal norms influence behaviors in different ways. Thus, it is often difficult to separate personal and social pathways.

Similarly, social capital can be considered as an attribute of an individual as well as of a community. As an individual, I have connections with individuals I trust—social capital—that I can draw on when I need information or seek help, such as recruiting volunteers for our biannual Ithaca City Cemetery cleanup. I often use my social capital to help students—for example, the other day I learned one of our Mexican online students was working on Colorado River issues, and I "eIntroduced" her to a colleague from Arizona who researches the Colorado River. Should these two individuals and their colleagues start working together, they may form trust, connections, and norms, or social capital. Such social capital in turn enables them to engage in citizenship and group stewardship behaviors (Ahn and Ostrom 2008; Krasny and Tidball 2015; see also chapter 5).

Positive youth development (chapter 14) and health and well-being (chapter 15) provide additional examples of the overlap between personal and social pathways. Youth develop assets such as leadership and communication skills through participating in an environmental education program with other students and community members. We can contribute to the well-being of individual program participants and their communities by providing opportunities to spend time in and steward nature and through building citizenship skills.

Finally, structural pathways are physical, economic, and organizational enablers of behaviors. The availability of fruits and vegetables or clean tap water, the accessibility of public transportation and recycling bins, the presence of a bike share program, traffic congestion fees in city centers, and tax rebates for solar panels and electric vehicles are all structural factors. Structural pathways address the structural barriers that limit an individual or a group's ability to engage in environmental behaviors.

Some may view environmental education as focused on a narrow set of personal pathways, in particular knowledge and attitudes, which have limited impact on behaviors (Hungerford and Volk 1990; Kollmuss and Agyeman 2002; Chawla and Cushing 2007). Yet environmental educators impact multiple personal pathways, through influencing participants' nature connectedness, self-efficacy, identity, personal norms, and sense of hope. Environmental educators also influence social pathways—both among their participants and, through their participants, the broader community. For example, participants develop trust while working together on a tree-planting project (Krasny, Kalbacker, et al. 2013) or conquering a group physical challenge in an outdoor program (Ardoin, DiGiano, O'Connor, and Podkul 2016). They then involve community members by organizing a community hike or environmental fair; we would expect these activities in turn to help build social capital (Ahn and Ostrom 2003; see chapter 13). Further, by planting trees, participants improve their physical environment; providing shady pleasant places for people to gather has been shown to increase social capital (Kuo et al. 1998; Krasny, Kalbacker, et al. 2013; Holtan et al. 2014; Krasny and Tidball 2015). In this way, stewardship, a social pathway, interacts with green infrastructure (e.g., urban trees), a structural pathway. Finally, environmental education participants can influence structural pathways through seeking knowledge about environmental issues and then advocating for particular solutions.

By the nature of their work, environmental educators are systems thinkers. They influence multiple factors within one program, and often these factors interact with one another. A program sets a social norm for a particular type of behavior—let's say vegetarianism—which in turn builds an environmental (vegetarian) identity among participants (see chapter 11). This identity reinforces the social norm and influences the personal norms of newcomers to the program (see chapter 12). Further, the intertwined pathway from identity and norms to behaviors is not always linear. Environmental behaviors can influence identity, norms, and social capital, which can in turn foster new behaviors. Thus, pathways to reach environmental education outcomes are complex; they are often interwoven and involve multiple feedbacks (figure 4.1).

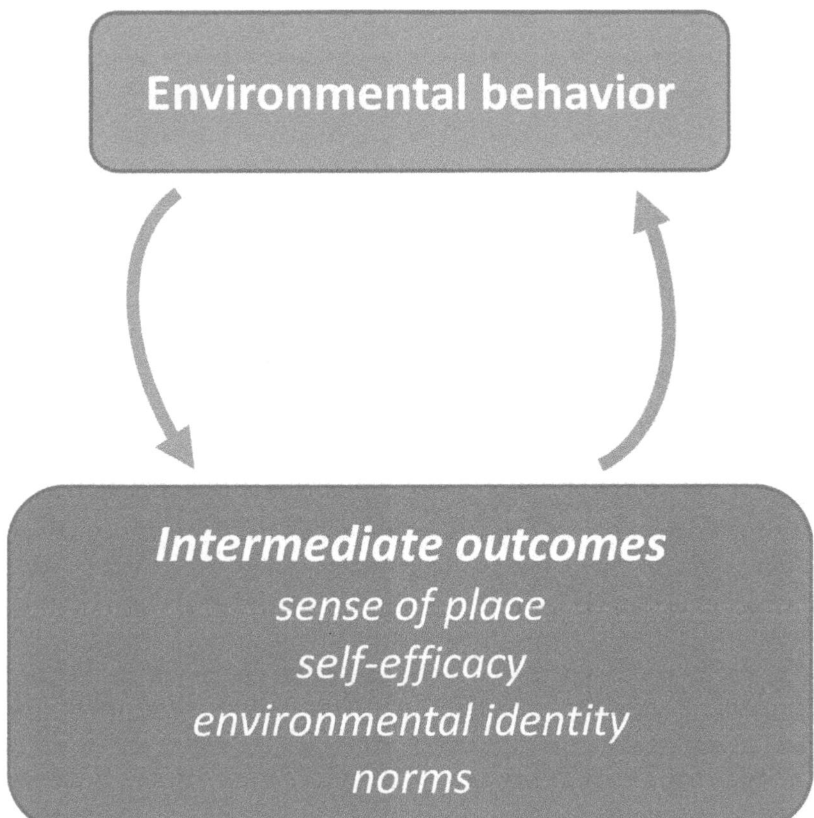

FIGURE 4.1. Feedback pathway. Intermediate outcomes, like sense of place, self-efficacy, environmental identity, and norms, can result from and lead to environmental behaviors.

Assessing Environmental Behaviors

Before assessing changes in environmental behaviors, environmental educators should ask themselves several questions. Do I want to focus on a single or multiple behaviors targeted by my program? Would I like to know about spillover behaviors? Should I prioritize behaviors that have the largest impact on the environment? And do I want to measure actual behaviors, or the impact of behaviors on reducing greenhouse gases, increasing area of land that is managed sustainably, or other environmental outcomes (Gatersleben 2013)? Finally, might there be a way to embed the assessment into my program activities to make the assessment fun and meaningful for participants?

To measure behaviors, evaluators often use self-report surveys that list any number of targeted and/or impactful behaviors, and ask respondents to indicate how often they perform those behaviors (Larson et al. 2015; see appendix). Although this is a straightforward and inexpensive means of assessment, it can be challenging to design a survey that measures behaviors that participants actually engage in. For example, surveys may ask about recycling plastic bottles or aluminum cans, but what if there is no recycling program where the respondent lives? Or surveys may ask about gas-saving car driving behavior (not speeding up and slowing down quickly), but what if the respondent has chosen a car-free lifestyle? One means to target behaviors relevant to your participants is to interview them to find out all the types of environmental behaviors they engage in, and then develop a survey based on their responses (Larson et al. 2015).

Self-report surveys are subject to reporting errors, such as when respondents exaggerate what they actually do, perhaps because they intend to do more or because they would like to please the evaluator (Gatersleben 2013). To address problems with self-report surveys, some assessments use an outsider (such as a roommate) to observe and record participants' actual behaviors (Chao and Lam 2011).

Another form of assessment is asking participants to keep a diary of their environmental behaviors, including by using apps designed to record environmental and eating behaviors (Mak 2015). These apps could be incorporated into programs as a fun activity that encourages environmental behaviors, and thus serve as an embedded assessment. For example, the JouleBug app (2018), which encourages sustainability behaviors through awarding points and allowing users to share their actions and compete with friends and other organizations, could be used to track such behaviors. Other apps are available for tracking daily eating habits, which could be used to assess reductions in meat and dairy consumption. For those concerned about litter, Litterati (2017) is a global online community that encourages participants to identify, map, and pick up litter. New apps are coming online—such as one being developed by Stanford University to track physical fitness and environmental behaviors (Landay and Crum, n.d.)—providing multiple opportunities for integrating assessment into program activities.

Some apps or computer interfaces directly connect with building-energy and water-use meters, which can provide a direct measure of the environmental impact of behavior change (Petersen et al. 2017). For public sphere behaviors, one could describe resultant policy changes, such as changes in the school cafeteria's number of meat-containing meals, number of single-use plastics, and kilograms of food waste.

5

COLLECTIVE ENVIRONMENTAL ACTION

> **[Environmental action] reflects citizenship traditions of participatory democracy, public work, and social justice because it includes youth directly in democratic processes, involves collective action toward some public purpose (e.g., creating a community garden), and ideally addresses the root causes of problems.**
>
> (Schusler et al. 2009, 123)

Highlights

- Collective environmental action includes a range of practices in which people engage in the civic and political life of their community or work with others to collectively steward green space.
- Collective environmental action is important because of its potential to have broader environmental and community impacts relative to individual lifestyle environmental behaviors, and to help program participants gain skills needed for lifelong contributions to the civic life of their community.
- Social capital and youth developmental assets create capacity to engage in collective environmental action.
- Environmental education engages participants in collective environmental action through a sequence of structured activities from community inventories to decision making to advocacy, as well as through stewardship programs in which participants engage in community gardening, coastal restoration, and other civic ecology practices.

Collective environmental action is a process whereby youth and adults create environmental and social change while building their capabilities for future civic participation (Schusler et al. 2009). In contrast to programs aimed at changing individual lifestyle behaviors like recycling or saving water, environmental action engages participants in planning for and taking collective action to address environmental problems, including their root causes. Environmental action also includes civic ecology practices, or hands-on stewardship and restoration activities conducted alongside other community members (Krasny and Tidball 2015; Krasny 2018).

Citizenship and Collective Stewardship

Collective environmental action encompasses citizenship behaviors, such as engaging in protests and advocacy, as well as stewardship practices such as volunteer litter cleanups or community gardening (Schusler and Krasny 2010; Alisat and Riemer 2015; Larson et al. 2015; table 5.1). Citizenship behaviors (also referred to as civic engagement) can take on a variety of forms, all of which attempt to influence democratic processes (Checkoway and Aldana 2013). *Grassroots organizing* focuses on social or environmental justice and often involves coalitions that challenge authority through protests and through public theater, dance, and other performances. For example, an urban agriculture nonprofit in New York City works with other grassroots organizations to engage youth in protesting against food insecurity (Delia and Krasny 2018); a grassroots group in Spain seeks to empower local residents through social theater focused on water conflicts (Jimenez-Aceituno et al. 2016); and the Trash Dance project in Texas empowered garbage workers by helping them choreograph their trash collection movements and those of their trucks, which they then performed in front of an audience of several thousand (Orr 2012).

A second form of citizenship behavior, *participation in political and government institutions*, encompasses serving on city commissions and speaking at public hearings or other public events (Schusler and Krasny 2010; Checkoway and Aldana 2013). For example, youth engaged in Earth Force's Community Action and Problem-Solving Process conduct research on environmental policies and present project proposals to public officials (Earth Force, n.d.), and youth in a climate education program in Michigan presented what they had learned to fellow students, teachers, and parents (Stapleton 2015).

A third form of citizenship behavior, *intergroup dialogue*, brings together distinct identity groups to discuss their differences and to foster collaboration across identities (Checkoway and Aldana 2013). In Environmental Issues Forums and Climate Courage Resilience Circles, people discuss and devise hopeful actions to address climate change (Armstrong et al. 2018; NAAEE 2018a).

Finally, a fourth type of citizenship behavior, *sociopolitical development*, focuses specifically on engaging youth of color and low-income youth in addressing

TABLE 5.1 Collective environmental actions

COLLECTIVE ACTION	DEFINITION	EXAMPLES
Citizenship (also called citizen engagement)	Group actions taken to influence policy (Stern 2000; Checkoway and Aldana 2013; Larson et al. 2015)	See each type of citizenship action below.
Grassroots organizing	Organizing to generate power and influence decisions of established institutions (Checkoway and Aldana 2013)	Protests, public theater, dance, and other performances
Citizen participation	Participation in established political and governmental institutions (Checkoway and Aldana 2013)	Conducting research on environmental policies and presenting policy proposals to public officials, speaking at public hearings or other public events, serving on advisory groups
Intergroup dialogue	Discussion of differences and fostering collaborative action across distinct identity groups (Checkoway and Aldana 2013)	Environmental Issues Forums and Climate Courage Resilience Circles that bring together people to deliberate and offer hopeful actions related to climate change (Armstrong et al. 2018; NAAEE 2018a)
Sociopolitical development	Social and political development of marginalized youth (e.g., ethnic minorities) to strengthen their ability to take collective action to address injustice (Checkoway and Aldana 2013). Similar to Social Justice Youth Development (Ginwright and Cammarota 2002) or Critical Pedagogy of Place (Gruenewald 2003; see chapters 9 and 14)	Training young people to become community organizers around issues of access to green space and industrial and traffic pollution in the Bronx (McKenzie et al. 2017)
Collective stewardship	Collectively managing a shared resource, like a plot of land or a stream, to improve habitat or biodiversity and to provide ecosystem services (Larson et al. 2015).	Creating bioswale gardens to reduce runoff, planting milkweed to improve monarch butterfly habitat, other civic ecology practices (Krasny and Tidball 2015; Krasny and Snyder 2016; Krasny 2018)

injustice (Checkoway and Aldana 2013). This type of engagement aligns closely with social justice youth development (see chapter 14). Examples come from Youth Ministries for Peace and Justice, which trains young people to become community organizers around issues of access to green space and industrial and traffic pollution in the Bronx (McKenzie et al. 2017), and a program to address the problem of children in South African townships playing with hazardous medical waste by training home health care workers in proper waste disposal (Krasny, Mukute, et al. 2017).

Collective *stewardship* refers to a group of people voluntarily working together to manage a shared resource, like a plot of land or a stream, to improve habitat or biodiversity and to provide ecosystem services. Activities include growing food in a church donation garden, creating bioswale gardens to reduce runoff from city streets, installing artificial oyster reefs to protect shorelines and restore oyster populations, reestablishing native species in parks, restoring mangrove forests or village groves, litter cleanups, and other civic ecology practices (Meinzen-Dick et al. 2004; Krasny, Lundholm, et al. 2013; Krasny and Tidball 2015; Larson et al. 2015; Krasny 2018). Communities that come together to engage in collective stewardship often already have established trust, social connections, and prosocial norms—or social capital. Engaging in collective stewardship action can also create social capital (see chapter 13).

Author Reflections

My first visit to a community garden—the Open Road Community Garden in the Lower East Side of Manhattan—has had a profound impact on my career for over twenty years. I saw how two groups—youth attending a school bordering the east side of the garden, and Bangladeshi immigrants attending a mosque on the garden's west side—could come together to experience the healing benefits of stewarding nature and community, and to learn from each other about the planting and cultural practices of their "place" of origin—places like the southeastern US, the Caribbean, and Bangladesh. In so doing, they were creating new "places" that brought together community and nature in New York City. After seeing similar practices in other cities, among refugees, and after disaster, my colleague Keith Tidball and I coined the term "civic ecology." In addition to community gardening, civic ecology encompasses other collective stewardship practices such as volunteer tree planting, mangrove and oyster restoration, and litter cleanups. In our book, *Civic Ecology: Adaptation and Transformation from the Ground Up* (Krasny and Tidball 2015), Keith Tidball and I explore a multitude of civic ecology practices and their outcomes for individuals, communities, and the environment.

No matter what city or neighborhood I travel to, I have always felt welcomed when I venture into a community garden. I have volunteered for an invasive species removal day at a park in Miami, participate regularly in cleanup activities sponsored by Friends of Ithaca City Cemetery, and cofounded the Cornell student club Friends of the Gorge to help steward our campus natural areas. But I have continually been troubled by the small scale of these efforts in light of the seemingly overwhelming social and environmental problems we face. Recently, I have discovered the many connections between hands-on community stewardship and citizenship behaviors that could have a larger impact. One example comes from the group "the Ugly Indian" (TUI 2010), which plans its volunteer "spotfixes" to convert trashed-strewn lots to pocket parks near municipal buildings. They see their spotfixes as a form of gentle protest to nudge government officials into taking responsibility for clean open spaces in Indian cities (Abhyankar and Krasny 2018). Authors in the book *Grassroots to Global: Broader Impacts of Civic Ecology* (Krasny 2018) explore this and other pathways by which civic ecology practices link with advocacy and similar citizenship behaviors to have impacts beyond a single patch of land or water.

Both citizen behaviors and collective stewardship are often combined with data collection. Such "public participation in scientific research" (Shirk et al. 2012) ranges from local participatory action research (Mordock and Krasny 2001) to global citizen science projects (Dickinson and Bonney 2012). Methods encompass interviews with community members to identify an environmental issue of local concern, monitoring the impact of habitat enhancement on bird or insect populations, and measuring the social connections formed among youth and adults working in a community garden (Kyle and Kearns 2018; Silva and Ramirez 2018; Earth Force, n.d.).

Opportunities to participate in collective environmental action vary depending on where you live. For example, protests and voting are possible in Western democracies but might be restricted and pose significant personal risk in other countries. In some countries, political corruption and lack of trust, alongside lack of meaningful opportunities to influence policy, limit people's motivation to engage in community action. In this situation, volunteering for a civic ecology practice or taking action within a school or university may provide alternatives to public protest and advocacy (Crocetti et al. 2012). For example, student members of a university environmental club in Beijing worked with campus cafeteria workers to reduce onetime plastics use (Yu 2018a).

For youth of color and low-income youth in the United States, opportunities to participate in environmental action may be constrained by lack of civic engagement in school and after-school club programs, lower rates of college attendance where service learning is often emphasized, and fewer adult role models owing to high ratios of children to adults and a significant proportion of men serving jail time (Flanagan and Levine 2010). Although minority participation in mainstream environmental organizations has historically been low (Taylor 2015, 2016), it is generally high in community gardening, tree planting, and other urban civic ecology practices (Musick et al. 2000; Saldivar and Krasny 2004; Eizenberg 2013; Fisher et al. 2015; Reynolds and Cohen 2016; Krasny 2018), as well as in community action, faith-based, and political groups (Sundeen 1992; Musick et al. 2000; Sundeen et al. 2009; Kyle and Kearns 2018). Thus, environmental action may serve as an approach to environmental education that addresses justice and other concerns in low-income communities and communities of color.

Why Is Collective Environmental Action Important?

> **Environmental education practice in the USA often focuses on promoting personal responsibility and environmentally conscious individual lifestyle choices. However, it does not always adequately address the economic and political structures that limit the freedom of individuals to make those choices.**
>
> (Schusler and Krasny 2008, 268–269)

Collective environmental action can have broad impacts on the environment, communities, and individual participants through multiple pathways (Riemer et al. 2016).

- Through protests and advocacy, participants in environmental action address structural and social justice issues that are the root causes of environmental problems (Checkoway 2012; Alisat and Riemer 2015; Krasny, Mukute, et al. 2017; McKenzie et al. 2017; Delia and Krasny

2018) and may even help spur institutional change and environmental movements (Witt et al. 2018).

- Through stewardship projects in public spaces visible to community members and government officials, environmental action can engage new participants and influence government policy (Abhyankar and Krasny 2018).
- Through working together, participants in environmental action programs create and reinforce existing social norms and social capital, thus building capacity for future environmental action and impacts (Kassam et al. 2018; see also chapters 12 and 13).
- Participating in environmental action helps youth develop assets that enable them to engage in future, more effective action. These assets include citizenship competence, self-confidence, agency, system knowledge, open-mindedness, ability to interact positively with others, and communication, critical thinking, and coping skills (Volk and Cheak 2003; Schusler et al. 2009; Schusler and Krasny 2010; Riemer et al. 2014; Schusler and Krasny 2014; Delia and Krasny 2018; see also chapter 14).
- Environmental and civic engagement can help fulfill youths' need to belong and to feel a sense of purpose, and can build social networks that enable youth to access educational and job opportunities (Flanagan and Levine 2010; Fisher et al. 2015; Delia and Krasny 2018).

In short, environmental action builds both community and individual capacity, which creates the conditions for further civic engagement and collective action. This type of feedback, where youth and adults influence their surroundings, thus creating greater opportunities for themselves and future environmental action, is not generally possible through consumer, energy, and similar individual lifestyle environmental behaviors (figure 5.1; see also chapter 4; Schusler et al. 2009; Wilkenfeld et al. 2010).

Pathways to Collective Environmental Action

After-school, summer, and intern programs are important training grounds for youth civic engagement. In contrast, civic education in schools has had limited success in engaging youth, perhaps because civics classes that focus on facts such as the branches of government, and that lack vigorous discussion of relevant issues, fail to motivate students (Checkoway 2012). In after-school and summer programs hosted by community and environmental organizations, youth participate in grassroots organizing and respectful dialogue about environmental and social justice issues, and thus learn through real-life experiences (Checkoway

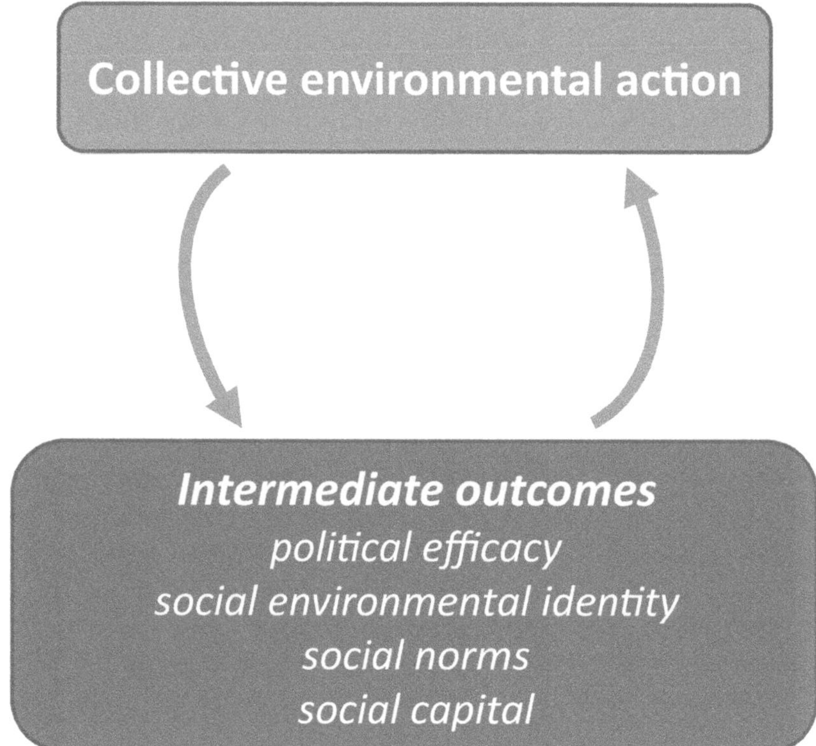

FIGURE 5.1. Feedback pathway. Intermediate outcomes, like political efficacy, social environmental identity, social norms, and social capital, can result from and lead to collective environmental action.

and Aldana 2013). Educators in these organizations constantly balance letting youth make their own decisions and assume leadership roles, with scaffolding and mentoring youth who lack the experience to take on these responsibilities (Schusler et al. 2017; see also chapter 14). When youth achieve success in environmental action, they gain civic and other forms of efficacy, which are precursors to future environmental action (see chapter 10).

Although many have expressed concern about the negative impact of the internet on constructive dialogue, studies have shown that specific uses of the internet, such as seeking out information and engaging in respectful online dialogue, may in fact foster civic and political engagement (Shah et al. 2005; Bakker and Vreese 2011; Samsuddin et al. 2016). Further, the internet and social media are transforming the ways people engage in environmental and other types of action, for example through offering opportunities to shape and personalize movement messages (Bennett and Segerberg 2013) and to mobilize and provide recognition to groups that engage in action (Abhyankar and Krasny 2018). This has led to people's social identity being shaped less by neighborhood, school, or work, "but rather by the manner in which they participate and interact through the social networks which they themselves have had a significant part in constructing" (Loader et al. 2014, 143). Thoughtfully mediated social media groups can also facilitate dialogue on contentious issues such as how to address climate change. Further, youth and adults engaged in face-to-face environmental actions may share their experience and perspectives through Facebook and other social media (Abhyankar and Krasny 2018). Thus, just as community organizations are vitally important in fostering civic engagement among diverse audiences, so too are social media and information seeking via the internet.

Collective action to steward community green space, streams, and other common-pool resources is more likely to occur in communities with high levels of social capital (see chapter 13). Social capital, in turn, is fostered through frequent communication that allows community members to gauge one another's trustworthiness and reactions to their ideas and actions (Ostrom 1990; Dietz et al. 2003). The ability of participants to define who participates is also important for successful collective stewardship—for example, limiting access to plots in a community garden to those who are committed to their fair share of weeding and garden upkeep. Other factors that foster collective action include community members who support the group rules and norms, are amenable to sanctioning those who refuse to comply, and have strategies to address conflicts should they arise (Ostrom 1990; Dietz et al. 2003). In short, trust, social connections, and demonstrated willingness-to-reciprocate behaviors that promote the collective good are present among groups that successfully steward natural resources (Ostrom 2010a).

Collective Environmental Action in Environmental Education

When youth take action to effect change, they can acquire skills related to planning, public speaking, fundraising, and organizing community support, as well as learn about civic-related concepts such as public purpose and power. Regardless of whether or not their

> **efforts are successful, engaging in collective action enables youth to think critically about the kind of world they want to live in. It also can enhance their understanding of social, economic, and political systems as they identify opportunities for and obstacles to realizing their vision.**
>
> (Schusler and Krasny 2008, 273)

Environmental action programs draw on a long-standing tradition of democratic participation in environmental education (Hart 1992; Mordock and Krasny 2001; Reid et al. 2008; Læssøe and Krasny 2013), including pedagogies such as action competence (Jensen and Schnack 1997) and programs such as Earth Force (Mueller-Sims 2016) and Growing Up in Cities (Chawla 2001). Participatory approaches emphasize the role of youth in deciding the environmental issue they will work on, planning how to address that issue, and taking action. In contrast, some environmental action programs engage youth in ongoing hands-on planting and other activities where they learn from elders and see immediate results of their work, but have limited ability to make decisions (Krasny, Lundholm, et al. 2013). Ideally all programs include critical reflection where participants discuss the broader meanings and implications of their work.

Before embarking on a civic engagement or collective action program that involves community activities and partnerships, educators will want to consider the context in which their programs will take place, including the constraints and opportunities for action. Because environmental action programs generally involve interactions with local government, they may be constrained by government policies toward protests, advocacy, and public engagement. Knowing the risks of certain types of actions, and which individuals in positions of power are willing to listen to participants' perspectives and are interested in addressing environmental issues, is important. When environmental action programs are conducted in collaboration with nonprofit community and environmental organizations, knowing the organizations' capacity, trustworthiness, reputation, and interest in working with participants is crucial. In addition, educators will want to consider the past experiences of their program participants. For example, youth growing up in impoverished neighborhoods or societies with rampant corruption may have low individual and community "outcome expectations"—that is, the expectation, gained through past participation in or observation of successful community action, that one's actions will make a difference (Chung and Probert 2011). Youth and community members with low outcome expectations may need more time to build trust and more guidance in interacting with others and in taking on responsibility at various stages of an environmental action program.

The nonprofit Earth Force uses the six-step Community Action and Problem-Solving Process to engage youth in environmental action. During the first step, youth conduct a community environmental inventory using websites, popular media, interviews, maps, walking tours, and other sources. Students use this information to decide what environmental issue to focus on (step 2). During the third step, youth explore existing policies and practices related to the issue, which leads to step 4, where they narrow down their list of project goals and change strategies. After writing their action plan and taking action to address their goals (step 5), students reflect on what they learned, what impact they had, and how their work can inform future projects (step 6) (Earth Force, n.d.). Likewise, youth in Growing Up in Cities programs document priorities for environmental action through mapping their neighborhood and conducting interviews, and then vote on a course of action. They work with adults experienced in community organizing, and local politicians help them understand how to navigate the policy process (Chawla and Cushing 2007). Similarly, action competence pedagogy outlines how youth develop the competence needed for lifelong action, through engaging in planning and through taking environmental actions focused on changing policy to address root causes of problems (Jensen and Schnack 1997; Morgensen and Schnack 2010).

Civic ecology education, or young people and other novices conducting hands-on environmental stewardship alongside more experienced adults, provides a pathway for engaging youth in collective action to steward a commonly held resource like a city park (Krasny and Tidball 2009a; Krasny et al. 2009; Krasny, Lundholm, et al. 2013). For example, the Cornell University Garden Mosaics program involved urban youth in working alongside and learning from elder community gardeners, and thus engaging in collective action around stewarding public green space in cities (Krasny et al. 2005; Krasny and Tidball 2009b). Because youth in civic ecology programs engage directly with ongoing community environmental stewardship groups, they do not generally become involved in the planning and decision-making processes inherent to Earth Force, Growing Up in Cities, and similar policy-focused environmental action programs.

Partnering with a natural resources manager or landscape architect can make native plant restoration, tree planting, or other stewardship practices more effective. Regardless of the stewardship practice, programs should consider an adaptive management strategy (Armitage et al. 2009; Plummer 2009), wherein participants monitor any outcomes from their actions (e.g., street tree survival), consider how well they are meeting their objectives, and adapt their practices accordingly.

Participants in all forms of environmental action will benefit from critical reflection about their collective work. Guided discussions of ideas about democracy, injustice, responsibility, resource management, and related concepts can help youth to reflect on and draw meaning from their actions, consider the broader implications of their work, and situate their work in the larger public sphere (Schusler et al. 2009; Delia 2013; Smith et al. 2015).

Finally, educators should consider the duration and intensity of their environmental action programs, as well as their role as mentors and role models (Chawla and Cushing 2007; Riemer et al. 2014). They may need to balance their authority with opportunities for youth autonomy (Schusler et al. 2017; see also chapter 14). Further, social connections formed with peers and adults during environmental action programs are needed to sustain such action into the future (Riemer et al. 2016).

Assessing Collective Environmental Action

Assessment can focus on environmental, community, and individual outcomes of collective environmental action. For stewardship projects, one might measure environmental impacts using kilograms of litter collected, area of land cleared of invasive species, number of trees planted, or area of vacant lot converted to a community garden (Krasny, Russ, et al. 2013). Citizen science monitoring protocols, such as those of the Great Pollinator Project (AMNH 2012), Project FeederWatch (2010), and Monarch Watch (2014), may be used to measure changes in biodiversity. Environmental and social impacts of community gardens can be assessed with simple measures (e.g., weight of vegetables, short happiness survey) described in Farming Concrete's toolkit (Design for Public Space 2015). For program participants, environmental educators can use measures of civic efficacy, social capital, and other intermediate outcomes covered in the chapters of this book. Finally, educators wishing to know how often participants conduct environmental actions after their programs can use the Environmental Action Scale (Alisat and Riemer 2015; see appendix). Survey questions cover types of civic engagement (Checkoway and Aldana 2013), including grassroots organizing (e.g., Within the last six months, how often, if at all, have you worked with an environmental justice group?); civic participation in political and government institutions (. . . took part in a protest around an environmental issue?); intergroup dialogue (. . . used online tools, such as Facebook or YouTube, to raise awareness about environmental issues?); sociopolitical development (. . . volunteered or worked with an environmental group or political party?); and environmental stewardship (. . . participated in nature conservation efforts, such as planting trees, watershed restoration?).

Because the types of environmental actions participants can engage in vary according to the political, cultural, and environmental context, assessments should be adapted for particular settings. Further, educators can use student reports and other program artifacts (e.g., letters to city council) as evidence of environmental action and its outcomes.

Part III

INTERMEDIATE OUTCOMES

6

KNOWLEDGE AND THINKING

Highlights

- Knowledge about the environment and environmental problems does not generally lead directly to environmental behaviors.
- System knowledge is knowledge about the environment, action-related knowledge is about actions to address environmental problems, and effectiveness knowledge is about which actions are most effective in addressing those problems.
- System knowledge can impact action-related and effectiveness knowledge, which in turn lead to pro-environmental behaviors.
- Systems thinking entails understanding system structures, dynamics, and functions, and that changing one aspect of a system can have unintended consequences.
- Critical thinking involves assessing the credibility of sources of information, being able to reach decisions based on credible information, and being open to changing one's decisions as new information emerges.
- Systems and critical thinking can help students make informed decisions about environmental issues.
- Whereas short-term, student-centered activities can enhance system knowledge, longer-term student-led investigations foster systems and critical thinking.

Study after study has shown that learning about the environment and environmental problems does not generally mean a person will adopt environmental behaviors. Several factors explain this disconnect between knowledge and action. For example, I might know about the science of climate change and why changing climate is a threat, but not know what I can do about it. I may even know about several potential actions I can take, but not know which ones will be most effective. Or I may belong to a group, as do many political conservatives in the United States, that denies climate change, and my identity as a conservative is a barrier to taking action. For these reasons, environmental educators who wish to change behaviors need to consider what types of knowledge they convey, and to link their efforts to build knowledge with strategies to build efficacy (see chapter 10), environmental identity (see chapter 11), and other intermediate outcomes.

Certain types of knowledge, alongside other factors like identity and efficacy, are more likely to lead to environmental behaviors and actions. In particular, knowledge about actions one can take and the efficacy of those actions is more likely to lead to environmental behaviors than knowledge about environmental systems. In this chapter, we cover three types of knowledge: system, action-related, and efficacy. Then we go beyond just knowledge—or what people know—to encompass how people think. Systems thinkers consider the complexity of human and natural systems and of environmental problems, whereas critical thinkers are able to revise their judgments and actions—including those targeting the environment—as new information becomes available.

What Are Environmental Knowledge, Systems Thinking, and Critical Thinking?

Different types of environmental knowledge and systems and critical thinking are crucial to environmental literacy. An environmentally literate person "both individually and together with others, makes informed decisions concerning the environment; is willing to act on these decisions to improve the well-being of other individuals, societies, and the global environment; and participates in civic life" (Hollweg et al. 2011, 2–3). Environmentally literate individuals have knowledge of ecological and social systems and of environmental issues (system knowledge) and of ways to address those issues (action-related knowledge). Environmentally literate individuals also have critical thinking skills, including the ability to use evidence to analyze, investigate, and make personal judgments about and plans to resolve environmental issues (Hollweg et al. 2011). A social justice or "critical ecological literacy" also entails individuals posing questions and taking action to address social and environmental injustice (Cermak 2012).

Knowledge

Three types of environmental knowledge are relevant in discussing environmental behaviors: system knowledge, action-related knowledge, and effectiveness knowledge (Frick et al. 2004). System knowledge is about ecosystem processes and environmental problems. For example, an environmental education program might teach about the water cycle and how farming practices impact the water cycle. Students might investigate whether producing a kilogram of almonds or a kilogram of chicken or beef requires more water. System knowledge is closely tied to systems thinking (see below).

Action-related knowledge is about what actions a person can take to address environmental problems (Frick et al. 2004). If people understand that growing beef or almonds requires a lot of water and that certain agricultural practices are negatively impacting the groundwater, they have a variety of options to address the problem, such as reducing consumption of these foods or trying to influence agricultural policy. Note that to understand and assess various options, knowledge of agricultural and groundwater systems (system knowledge) is required.

Effectiveness knowledge is understanding which behaviors and actions are most effective in addressing an environmental problem (Frick et al. 2004). Would eating chicken rather than beef or almonds help reduce agricultural water use? Is signing a petition, starting an environmental club, calling my political representative, joining a protest, or attending a town hall meeting more effective at addressing groundwater issues? Similar to action-related knowledge, effectiveness knowledge requires an understanding of environmental systems.

Systems Thinking

> **[A system is] an integrated whole whose essential properties cannot be reduced to those of its parts. [A system's properties] arise from the interactions and relationships between the parts.**
>
> (Capra and Luisi 2014, 10)

A systems thinker recognizes the multiple interconnections between natural and social components of a system, and that even slight changes to one component "can radically alter how the system will behave and adapt" (Randle 2014, 17). Systems thinkers apply their understanding of the interactions of parts of a system to making environmental decisions (Davis and Stroink 2016).

To help make informed decisions, systems thinkers learn to recognize three system attributes: structures, dynamics, and functions. Structures are individual components in a system, including the plants, animals, soils, and even built infrastructure like buildings and dams. System dynamics are changes over time in how the structures interact. Because they are difficult to observe and involve a series of complex processes where one thing leads to another and then another, system dynamics are considered the most difficult aspect of systems thinking. Rather than focus on individual organisms within a system, like honeybees, system dynamics directs attention to the patterns that emerge when many individuals interact, as in a beehive. Finally, system functions refer to the purposes or roles of the system's structures and dynamics. Functions include ecosystem services such as food, fiber, fuel, and filtering water runoff from streets and agricultural fields (Danish 2014; Garavito-Bermúdez et al. 2016; Hmelo-Silver et al. 2017).

Another component of systems thinking is understanding that systems have emergent properties that humans cannot control. In fact, much of the push toward systems thinking stems from observing the negative impacts of farming, forestry, and fisheries practices where humans assumed they could control an ecosystem to meet human needs. Farmers and foresters developed monocultures of single crops, and foresters suppressed forest fires for years. Although these practices have been successful in meeting the food and fiber needs of billions of people around the world, they also have led to the emergence of destructive insect outbreaks and massive forest fires (Folke et al. 2002; Davis and Stroink 2016). Humans have learned the lesson that trying to enhance productivity of one part of a system, while ignoring the dynamics among the system's myriad components, can lead to unintended consequences.

In short, systems thinkers recognize how complex and dynamic systems are and make decisions accordingly (Clark et al. 2017; Hmelo-Silver et al. 2017). They grapple with how humans impact system structures, dynamics, and functions, including by enhancing or negatively impacting the ecosystem services a system provides (table 6.1). Armed with such understandings, systems thinkers turn to critical thinking to explore ways to address environmental problems.

Critical Thinking

> **The ideal critical thinker is habitually inquisitive, well-informed, trustful of reason, open-minded, flexible, fair-minded in evaluation, honest in facing personal biases, prudent in making judgments, willing to reconsider, clear about issues, orderly in complex matters, diligent in seeking relevant information, reasonable in the selection of criteria, focused in inquiry, and persistent in seeking results which are as precise as the subject and the circumstances of inquiry permit.**
>
> (Facione 1990, 2)

Critical thinkers seek out multiple sources of information about a problem and how to address it. They are able to judge the credibility of information by asking

TABLE 6.1 Abilities of systems thinkers illustrated using a community garden example (adapted from Assaraf and Orion 2005)

ABILITY	COMMUNITY GARDEN EXAMPLE
1. Identify system structures	Structures: people, soils plants, insects, birds, water, fences, buildings
2. Identify system dynamics	Photosynthesis, nutrient cycling, evapotranspiration, feeding and predation, people connecting with each other and with nature
3. Organize the system's structures and dynamics within a framework of relationships	People plant and harvest vegetables; children and adults talk with each other and form social connections; plants absorb nutrients and water from the soil and transpire water into the atmosphere; insects pollinate plants; insects and birds feed on plants and other insects
4. Make generalizations	The community garden is an open system, which includes outside inputs (e.g., visitors and short-term volunteers, vegetable scraps from nearby restaurants that are made into compost).
5. Understand the hidden dimensions of the system	In soils contaminated by lead, roots can take up the lead and impact lead content in some plants. Children playing in the garden may absorb lead.
6. Understand the cyclic nature of systems	Bacteria and fungi in the soil decompose plant material, thus "recycling" nutrients for the plants.
7. Understand the past and make predictions about the future	Past use of the garden site for dumping caused soil contamination, created an eyesore in the neighborhood, and encouraged criminal activity. The presence of neighbors actively working together to transform the site to a community garden will discourage unwanted activities in the future.
8. Understand the relationship of the system to smaller and larger systems	The community garden consists of subsystems such as raised beds for planting vegetables. The garden is embedded in a neighborhood system that includes water, plants, people, streets, and buildings.

clarifying questions and by conducting their own investigations. Thus, they are open-minded yet well informed by credible, evidence-based information. Critical thinkers also are constantly on the lookout for new information to prove or disprove any conclusions they have made about a system or problem. Importantly, they make changes in their decisions and actions based on new information (Ennis 1993). Thus, critical thinkers are willing to reconsider and revise their views when the evidence and reflections suggest the need to change (Facione

TABLE 6.2 Critical-thinking skills using climate change example (Facione 1990)

SKILL	CLIMATE CHANGE EXAMPLE
Interpret. Categorize and describe the significance of a range of experiences, data, events, beliefs, and situations.	Describe significance of climate change to recent hurricanes and typhoons, floods, fires, and droughts.
Analyze. Identify problems and identify and analyze arguments about ways to address those problems.	Analyze arguments for ways to address climate change, including reducing individual energy and meat consumption, advocating for carbon tax policies or research on carbon capture technologies, or building seawalls and diverting rivers.
Evaluate. Assess the logic and validity of claims and arguments.	Use Drawdown (n.d.) and Energy Innovation (2018) resources to assess effectiveness of individual consumption behaviors and government policies. Use other online resources to investigate carbon capture and adaptation options (e.g., building seawall, diverting river).
Infer. Question evidence, suggest alternatives for resolving a problem, and decide the best course of action given current information.	Read background on how Drawdown solutions were prioritized, and decide on an individual behavior or policy action. Read other online sources and judge their credibility.
Explain. Clearly explain decisions and the reasoning behind those decisions.	Develop presentation on rationale behind choosing specific behaviors and actions.
Self-regulate. Reflect on reasoning and motivations and on biases that may have influenced that reasoning; design procedures to adjust when mistakes are made.	Reflect on biases, reasoning, and reactions from others to changes in behaviors and collective actions. Collect additional information as needed and make adjustments to increase effectiveness.

1990). Environmental education provides ample opportunities to help participants build critical thinking skills (table 6.2).

Why Are Knowledge, Systems Thinking, and Critical Thinking Important?

People who engage in systems thinking attend to and process system-related information more broadly and recognize complex

> **causal relationships and patterns of change; as a result, they are more likely to make decisions that enhance the well-being of the systems they interact within and depend on.**
>
> (Thibodeau et al. 2016, 753)

Research has demonstrated that knowledge alone is not sufficient—or in some situations even necessary—for people to adopt environmental behaviors. This may be especially true in places where people's strong political and cultural identities take precedence over the facts in determining their behaviors. For example, in the United States, Republicans have tended not to support climate policies, and Republicans with high levels of science literacy were even less likely to support climate policies (Kahan 2015).

Does this lack of connection between knowledge and behaviors suggest we should simply give up on "educating"? Despite the need to go beyond helping people gain knowledge about ecosystems and environmental problems, there are valid reasons to foster different types of knowledge alongside systems and critical thinking.

- System knowledge can be taught in K–12 classrooms, and thus can improve the science literacy of millions of children.
- System knowledge provides a foundation for action-related and effectiveness knowledge (Fremerey and Bogner 2014).
- Action-related knowledge predicts environmental behaviors; it is likely that effectiveness knowledge is also associated with environmental behaviors (Fremerey and Bogner 2014).
- Systems thinking addresses the complexity of natural and human-dominated systems, including how particular actions can have unintended consequences.
- Systems thinking entails realizing that humans are part of natural systems, which is related to nature connectedness (Davis and Stroink 2016; Thibodeau et al. 2016; see also chapter 8) and to self- and group efficacy in the context of addressing community problems (Clark et al. 2017; see also chapter 10); nature connectedness and efficacy predict environmental behaviors.
- Critical thinking is crucial to the ability to make informed decisions about ways to address environmental problems and to revisit and revise those decisions based on the constant flux of new information (Ernst and Monroe 2004).

How Do Knowledge and Systems and Critical Thinking Foster Environmental Behaviors?

The pathway from system knowledge to action requires multiple steps (figure 6.1). Individuals who have gained system knowledge about environmental problems may become interested in what actions they can take. Having been introduced to the range of possible individual behaviors and collective actions, individuals may next want to know which actions are most effective. In short, system knowledge does not lead directly to behavior change, but rather can impact action-related knowledge, which in turn is a strong predictor of behavior. Although researchers hypothesize that effectiveness knowledge also motivates environmental behaviors, such knowledge has not been widely available, and thus studies on its outcomes are limited (Frick et al. 2004; Roczen et al. 2014). This may be changing; organizations like Drawdown recently have compiled information on the relative effectiveness of one hundred actions to reduce greenhouse gases (Drawdown, n.d.), and Energy Innovations has an online tool to assess the impact of various energy policies on greenhouse gases (Energy Innovation 2018).

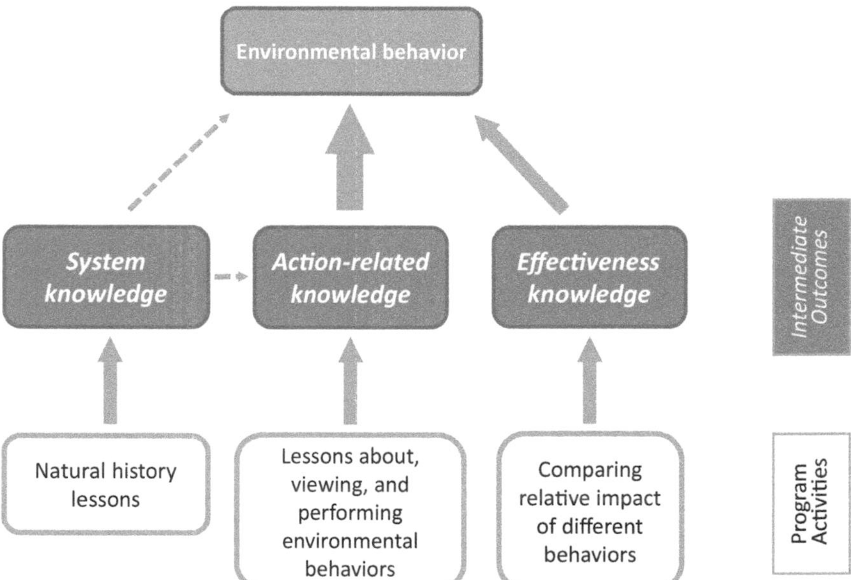

FIGURE 6.1. Knowledge pathways to environmental behavior. Action-related knowledge and effectiveness knowledge are more likely than system knowledge to lead to environmental behaviors.

Systems thinking has two potential pathways to environmental behaviors (Davis and Stroink 2016). First, people who understand that systems are complex and dynamic, and thus humans cannot control systems, may be more likely to make decisions and take actions that reflect this thinking. The leader of a community garden who is a systems thinker is aware of the dynamic interactions of plants, soil, and people within the garden, as well as the ways in which the community garden interacts with the surrounding neighborhood. Armed with these systems thinking skills, the leader might forgo hiring an outside firm to landscape the garden, even though the result would be attractive to outsiders. Further taking into account multiple interactions, the leader would encourage gardeners and neighbors to participate in formulating rules and choosing plants for the garden, thinking that such participation will lead to ownership in tending the garden and eventually to outside support. By considering multiple social-ecological interactions inside the garden and neighborhood "systems," the leader can make the garden a lasting neighborhood asset that provides multiple ecosystem services.

Just as community garden systems thinkers understand that the garden is part of the neighborhood, they also understand that humans are part of nature. Feelings of connectedness to nature have been shown to predict environmental behaviors (see chapter 8). Thus, nature connectedness is the second pathway from systems thinking to environmental behavior (figure 6.2; Davis and Stroink 2016; Thibodeau et al. 2016).

Similar to systems thinking, critical thinking aids individuals in making decisions about environmental behaviors and action. Importantly, critical thinkers continually seek out, analyze, and question the credibility of evidence and are willing to revise their actions based on what they discover. So a critical thinker who is managing a botanic garden education program might observe visitors learning about plants and climate change, and reflect on the fact that learning about the science is not providing the visitors with ways to act on this knowledge. He might then help the garden install solar panels and enroll in a smart energy program, and provide information on how visitors can similarly engage in actions that reduce greenhouse gases. The visitors, who are also critical thinkers, will want not only to change their energy consumption behaviors, but also to obtain knowledge about which behaviors have the greatest impact on reducing emissions, and will be open to adopting new strategies as more information becomes available. Because critical thinkers are able to judge the validity of information, they may also be less influenced by unsound judgments and claims, and base their decisions and actions on the best available information.

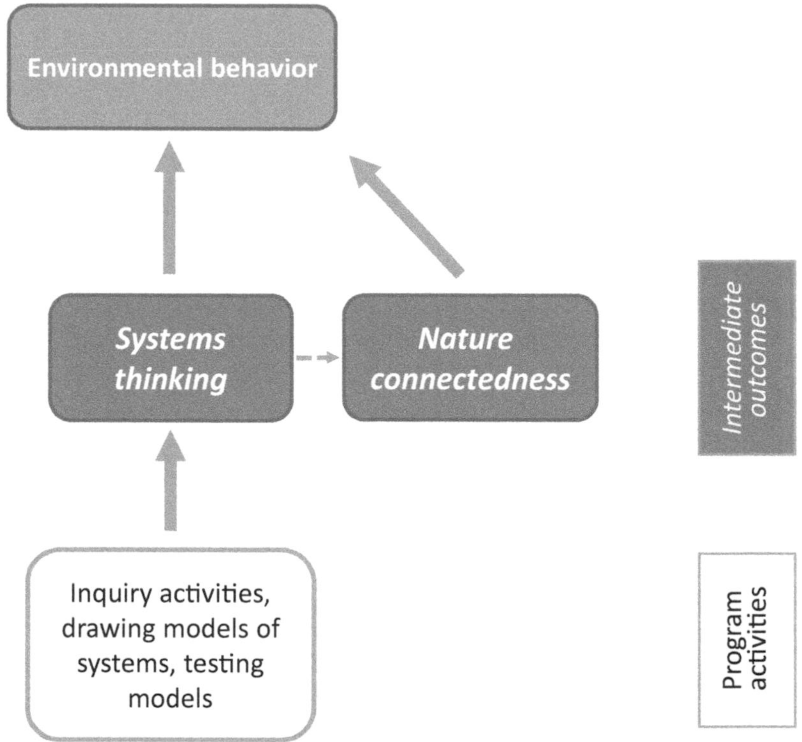

FIGURE 6.2. Systems thinking pathways to environmental behavior. Systems thinking can influence environmental behaviors directly and indirectly through fostering nature connectedness.

Author Reflections

One of my primary motivations for writing this book is my frustration with how hard it is to break out of the knowledge-attitudes-behavior paradigm of environmental education. Even though environmental education researchers have been challenging this paradigm with research-based evidence for over thirty years (Hines et al. 1986/87; Kollmuss and Agyeman 2002; Stern et al. 2014), even I find myself resorting to the paradigm. After all, I am an educator, so I often resort to teaching first, and only then question whether conveying knowledge will lead to the outcome I desire. Because of my work with community gardens and other civic ecology practices, I have come to realize that youth and adults may participate in a gardening, mangrove restoration, or tree planting project prior to learning about growing plants. Through engaging in these actions, they gain knowledge; thus behavior can precede knowledge, as it often does in everyday life.

In writing this chapter, I discovered a great source of effectiveness knowledge. Usually, when I click on an article about what I can do to save the planet, I see a long list, but no one tells me what is most effective. The nonprofit organization Drawdown has a list of one hundred actions to reduce greenhouse gases in order of effectiveness (Drawdown, n.d.). Because I constantly look for ways to reduce my environmental footprint, I was thrilled to learn how to prioritize my environmental behaviors. The three most effective actions I can take personally are reduce meat and dairy consumption, turn down the heat in my house, and restrict car and air travel. Turning off the lights and recycling are important, but nowhere near as effective as big energy consumers like meat production, heating systems, and flying. And if I am uncomfortable in lower temperatures, I can buy a personal heating device, like a foot-warming pad or rubber hot-water bag, to save energy (see chapter 3).

How Can Environmental Education Foster Environmental Knowledge and Systems and Critical Thinking?

> **Nature-protective behavior cannot be sufficiently explained using a pure rational/cognitive approach.**
>
> (Kals et al. 1999, 178)

Shorter-term interventions can foster knowledge gain. Longer-term programs, including those where students conduct well-designed investigations, can foster systems and critical thinking. Keep in mind that fostering knowledge and systems and critical thinking should be linked with other intermediate outcomes (e.g., nature connectedness) if the goal is to change environmental behaviors.

Knowledge

Carefully designed, collaborative, hands-on and investigative learning activities can enhance environmental or system knowledge. These activities often take place during school field trips; learning is enhanced when teachers prepare the students and follow up on the field experience in the classroom. Using real-life objects such as plant specimens, binoculars (Randler and Bogner 2009), and energy dashboard displays (Clark et al. 2017) also fosters system knowledge.

In one program shown to enhance system, action-related, and effectiveness knowledge about drinking water, groups of four students rotated between nine learning stations. At each station, students spent fifteen minutes reading, watching videos, and conducting small experiments before moving to the next station with different activities. After learning at the nine stations, students spent one hour visiting a nearby sewage treatment plant that used wetland plants to filter waste water. There they observed and performed small experiments on water purification (Fremerey and Bogner 2014). Similarly, students who conducted collaborative activities at learning stations during a one-day botanic garden education program enhanced their knowledge of climate change and plants; important to these knowledge gains were teachers preparing the students for what to expect prior to the field trip and helping students process what they had learned at the learning stations (Sellmann and Bogner 2013). Knowledge increases in a weeklong residential environmental education program at a national park were also attributed to hands-on collaborative learning, prior preparation of students to help them feel comfortable in the novel environment, and follow-up activities aimed at consolidating knowledge gains (Dieser and Bogner 2016). Interestingly, both live observations of bees coming and going from hives and watching videos of hive activity (virtual observations) increased students' system knowledge (Schönfelder and Bogner 2017). To build knowledge about system structures, dynamics, and functions, environmental educators can start with lessons on individual species and their adaptations and then widen the scope to include species interactions (e.g., food webs, human-nature interactions).

Students seeking to acquire action-related knowledge can use credible websites. In addition to Drawdown (Drawdown, n.d.), the Oceanic Society website has information on actions to reduce plastics pollution (Hutchinson 2014), the EnergySave website gives options for saving energy (EnergySave 2018), and the American Community Gardening Association and Cornell University have resources on starting and maintaining community and other learning gardens (ACGA 2018; Eames-Sheavly et al. 2018). Students can also pose questions about effective action—such as "What's worse: burning plastic trash, or letting it sit in a landfill?" on the "Ask Umbra" feature of Grist Magazine (Grist 2018).

Systems Thinking

A common approach used to foster systems thinking is students constructing and in some cases testing visual models of ecosystems or social-ecological systems. Constructing representations of systems and testing their ideas through modeling enables students to visualize each other's ideas and stimulates discussion (Hmelo-Silver et al. 2017). For example, participants might draw a model of the human recycling behavior system, showing how recycling behaviors are influenced by family and neighbors, city regulations, municipal waste services, and cultural values (Dittmer and Riemer 2012). After devastating floods in Colorado and New York, students who conducted hands-on restoration activities and collaboratively constructed models of the disaster "system" demonstrated systems thinking and action-related knowledge (Smith et al. 2015).

In a ten-week classroom project for kindergartners and first graders, students used software and drawing activities to learn about how honeybees collect nectar, and acted out bee nectar-collecting behavior. The children gained the ability to discuss bee structures (e.g., head, thorax), behaviors (nectar gathering), and functions (e.g., the honeybee dance aids in nectar collection for the group of bees), thus gaining systems thinking skills related to individual bees and the hive (Danish 2014). Although it may seem counterintuitive to start teaching about systems in kindergarten or first grade, middle school students who learned about systems in the elementary years more readily acquired systems thinking skills (Assaraf and Orion 2005). In sum, elements common to programs that achieve systems thinking include hands-on, inquiry-based, or stewardship activities and opportunities for integrating what has been learned through creating drawings of cycles (e.g., water cycle) or other visual models that include system components and their interactions (Assaraf and Orion 2005).

Critical Thinking

Environmental educators aiming to help learners acquire critical thinking skills can start by modeling such skills themselves, including demonstrating how they reflect on their biases, come to decisions through a process of logical reasoning, and revise their decisions and actions based on new information. Educators can explain the importance of critical thinking to students, and how emotions may be a barrier to such thinking. Investigation-based assignments can ask students to reflect on their biases; pose questions; analyze, critique, and discuss information, alternative perspectives, and arguments put forth in articles, videos, podcasts, and other media; and present and reflect on their own decision-making processes and actions (Ernst and Monroe 2004; Hofreiter et al. 2007).

Students in schools where the local environment serves as the context for learning across multiple disciplines may develop critical thinking skills, regardless of their achievement level. Such schools provide project- and issue-based learning experiences and use learner-centered and constructivist pedagogies

(Ernst and Monroe 2004). Programs conducted over multiple years allow students to practice their critical thinking skills in increasingly complex situations. Environmental educators working in nonformal settings (e.g., nature centers) can support school environment-based learning by offering sites for hands-on activities, mentoring student research, sharing their expertise, and providing teachers with professional development opportunities and access to funding and other resources (Ernst and Monroe 2004).

In one example, elementary schools in Hawaii used the Investigating and Evaluating Environmental Issues and Action curriculum across subjects. Working in small groups, fifth- and sixth-grade students chose a local issue to investigate, planned and conducted investigations, made recommendations to the community based on their results, and took actions to try to resolve the issue (Volk and Cheak 2003). According to a teacher,

> This 5th/6th grade configuration involves students in two years of issue investigation with its attendant skill development and application, and two years of planning and participating in the community symposium. During their fifth grade year, the students are "apprentices," as they learn the fundamentals of issue analysis, information accessing, instrument design, and data collection and interpretation—working in a group with one or more experienced sixth graders. During the sixth grade year, students assume the role of mentors and peer teachers as they lead their groups through the investigation. (Volk and Cheak 2003, 13)

The Investigating and Evaluating Environmental Issues and Action curriculum also was shown to increase critical thinking skills when implemented in a single subject area. For example, students in a social studies middle school class identified environmental issues, the relevant actors, and the actors' beliefs and values; selected an environmental issue for investigation; and developed research questions. They then sought out information from stakeholders and wrote persuasive essays and gave presentations based on their findings. This structured investigation was better able to foster critical thinking skills compared to environmental education programs lacking a series of activities to specifically build such skills (Robinson 2005).

In an environmental sciences course at a university in South Africa, students engaged in a series of structured steps to investigate citizen science projects. They started by posing the following question: "What are the possibilities?" This led students to discover different citizen science approaches from around the world, during which they became excited by the myriad of possibilities. Their excitement helped mitigate the pessimism that often comes with investigating environmental issues and set a tone whereby expressing diverse ideas and experiences was encouraged and respected. Next students asked, "What are the problems?" and investigated the challenges of one citizen science project. As they shared the results of these first two tasks, students created a question bank, which focused the next step in which they answered the question "What is known about citizen science?" Students discussed and compiled their observations into visual representations of what is known about citizen science, thus demonstrating systems thinking. For the assessment phase, students were asked, "Drawing on at least two examples [of student representations of what is known about citizen science], discuss the challenges and opportunities of citizen science for environmental monitoring." The professors summarized: "The Socratic method, interwoven with the explicit adoption of pluralistic thinking approaches which bridged the cognitive and affective domains, supported the students' development of critical thinking skills" (Belluigi and Cundill 2017, 966).

In sum, critical thinking interventions include structured, real-world investigations designed explicitly to teach critical thinking skills, as well as instructors who model the desired skills. Educators should also consider students' core values and even their emotions, as values and emotions can block our ability to think critically. Educators can begin by acknowledging students' core values and creating a safe environment for students to share different perspectives and experiences, leading to an expectation that differing perspectives are part of critical thinking. They can then move to discussions of evidence and fact-based information and logical arguments (Hofreiter et al. 2007; Belluigi and Cundill 2017).

Assessing Knowledge, Systems Thinking, and Critical Thinking

Short-answer tests or surveys are often used to measure knowledge and systems and critical thinking (Roczen et al. 2014; see appendix). You can also use embedded evaluation that links learning and assessment in your programs. For example, cognitive maps or drawings of systems created by participants as part of the learning activities can be used to assess system knowledge and thinking, and student-led presentations and discussions with community members can be used to assess critical thinking skills.

A cognitive mapping or drawing task for assessing system knowledge and thinking might ask students to diagram system components and draw arrows labeled with the interactions between various components. Students can add a brief description of the functions or ecosystem services provided by the system at the bottom of the drawing (Hmelo-Silver et al. 2017). If you use short-answer tests for system knowledge, decide whether you want students to acquire content specific to your program (e.g., a stream ecosystem or community garden social-ecological system) or more general knowledge across different systems. Similarly, for action-related knowledge, you can ask questions about how to reduce nutrients flowing into a stream or how to enhance plant growth and community engagement in a garden, or more generally about what actions people can take to reduce pollution or create access to urban green space.

To assess effectiveness knowledge, environmental educators can engage students in investigating the relative effectiveness of different behaviors. For example, participants in an energy use program can research government and NGO websites that detail what activities consume the most energy (e.g., heating, transportation). They can then focus on which home heating or transportation actions are available to them, and seek out information on their effectiveness in reducing energy use. Such investigations can serve as an embedded assessment of system, action-related, and effectiveness knowledge, for example, by asking students to describe local energy supply, energy use systems, and the effectiveness of energy reduction actions.

Educators can also use student-led investigations as embedded assessments of critical thinking. For example, students can be asked to evaluate the credibility of their sources of information, write reflective essays on the actions they decided to take, and present the results of their investigations at community meetings. Analysis of student presentations and written and digital media products serve as embedded assessment tools.

Asking students to choose one of several alternatives and explain their decision also can be used to assess critical thinking skills (TeachThought 2018). For example, a question about credibility of sources of information might ask students to choose from among several sources and defend their decision. To assess open-mindedness, students might write a short reflection, such as on an instance when they changed their position based on new information (Ennis 1993). Surveys are also available to assess systems and critical thinking skills (Ennis 1993; Ernst and Monroe 2004; Assaraf and Orion 2005; Davis and Stroink 2016; see appendix).

7

VALUES, BELIEFS, AND ATTITUDES

Highlights

- Values, like environmental protection and social justice, serve as broad guiding principles for multiple behaviors.
- Beliefs focus on what people think and have more limited impact on behaviors.
- Attitudes couple beliefs and emotions with a positive or negative judgment about an environmental issue or action.
- Environmental education programs can help children form environmental values, beliefs, and attitudes.
- Adults often have strongly held values, beliefs, and attitudes, which are difficult to change but may predict the level of support for environmental policies and should be considered in program planning.

Values, beliefs, and attitudes are all pieces of the puzzle of why people engage in environmental behaviors. Research results are mixed, with values, attitudes, and beliefs sometimes having strong relationships with behaviors and other times having little or no relationship (Stern 2000a; Heberlein 2012). Some researchers have gone so far as to suggest that instead of assuming attitudes are precursors to behaviors, environmental behaviors come first and attitudes follow (Eilam and Trop 2012). For younger audiences, environmental education may be one of multiple influences on values, beliefs, and attitudes, whereas for adults, often the best path is to design messages and programs around your audiences' already formed and often strongly held values, beliefs, and attitudes.

What Are Values, Beliefs, and Attitudes?

Values reflect the foundational principles that guide our lives. Beliefs are what we think but may or may not be factually true. Attitudes integrate beliefs, emotions, and judgments about an object like an animal or a behavior like conserving wildlife.

Values

Values are broad goals or principles that tend to be stable over time and that people find important in life—like equality, freedom, and environmental conservation. We use values to guide our behaviors across different contexts. Values may determine behaviors directly by helping us prioritize what is important to us or indirectly by providing a foundation for more specific beliefs, norms, and attitudes (Schwartz 1992; De Groot and Thøgersen 2013; Steg, Bolderdijk, et al. 2014).

Four types of values—biospheric, altruistic, egoistic, and hedonic—all play a role in environmental behaviors but in different ways. Biospheric values refer to a concern for nature and the environment without regard to human needs (De Groot and Thøgersen 2013). Although some think people who are wealthier and have basic needs met are more likely to hold biospheric values, people across different cultures, countries, and income levels value nature and the environment. Thus, biospheric values are thought to result from factors such as observing environmental degradation, directly depending on ecosystems for food, or even our genetic predispositions to "love" nature (Kellert and Wilson 1993; Brechin and Kempton 1994; Steg, Bolderdijk, et al. 2014). Because values provide broad guidance about what is right but do not identify specific "right" behaviors, people holding biospheric values may decide on different actions. For example, one person who values the environment might decide to go to the Galápagos Islands as an ecotourist, while another might decide to stay home because of concern about the carbon footprint of flying (De Groot and Thøgersen 2013).

Altruistic (or social-altruistic) values reflect a concern with the welfare of other human beings (Stern and Dietz 1994). People with altruistic values may act to protect the environment because they believe it will help others—for example, advocating for clean air will help children with asthma live healthier lives.

Egoistic values reflect concern for oneself, whereas hedonic values refer to individual pleasure. When environmental behaviors involve sacrificing comfort or pleasure, they work against these values. However, sometimes people's economic self-interest or status can be met through environmental behaviors. For example, installing solar panels may be cheaper than paying for fossil fuel energy and may be trendy in one's community. Similarly, the desire to have pleasurable experiences at times overlaps with environmental behaviors, such as when people enjoy planting a pollinator garden in their yard (Stern and Dietz 1994; Steg, Bolderdijk, et al. 2014).

In short, people may engage in environmental behaviors because they value helping the environment, the community, and even their family and themselves. However, only those holding biospheric values would likely hold environmental beliefs (Steg et al. 2011).

Beliefs

A belief is an acceptance that something is true, which may or may not be based on fact (Heberlein 2012). Just as I was writing this chapter, a US congressman expressed his belief that sea level rise is caused by boulders falling off cliffs into the ocean (which is definitely not true). Although both knowledge and beliefs are based on cognition, knowledge differs from beliefs because it is based on facts; for example, we know that increased ocean temperatures and water volume contribute to sea level rise. Whereas values are very broad, beliefs (and attitudes) focus on a particular "object" like sea level rise or particular behavior like saving energy (Heberlein 2012; De Groot and Thøgersen 2013).

An ecological worldview refers to one's beliefs about the relationships between humans and the environment. The New Environmental Paradigm (or revised New Ecological Paradigm, NEP) is commonly used to measure ecological worldviews (Dunlap and Van Liere 1978; Dunlap et al. 2000). It reflects beliefs about humans' ability to upset the balance of nature, the existence of limits to growth for human societies, and humans' right to dominate the rest of nature. The relationship between the NEP and environmental behaviors is generally weak compared to other measures such as values, attitudes about specific behaviors, and connectedness to nature (figure 7.1; Hines et al. 1986/87; Kaiser et al. 1999; Steg et al. 2011; Frantz and Mayer 2014), possibly because NEP measures general rather than specific beliefs and fails to capture an individual's direct, emotional experiences in nature (Mayer and Frantz 2004). Other environmental beliefs include those related to the consequences of environmental degradation for humans and the environment and one's responsibility for environmental degradation (Stern, Dietz, et al. 1995).

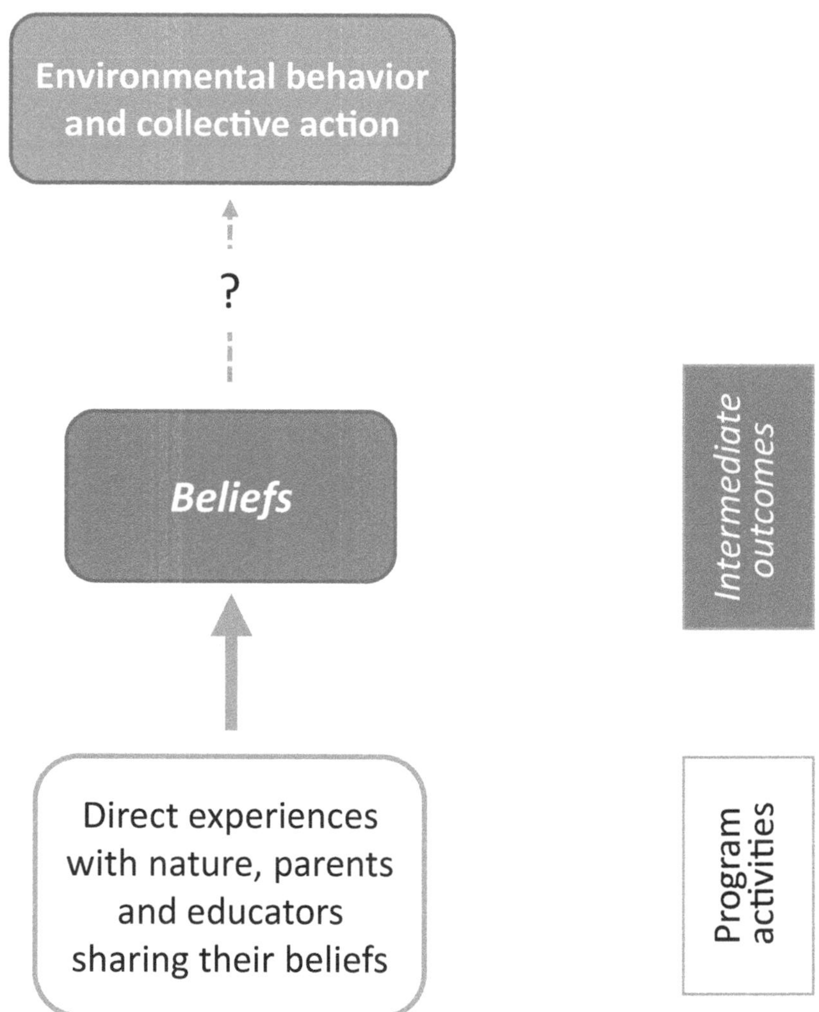

FIGURE 7.1. Environmental beliefs generally have a weak, if any, connection to environmental behaviors.

Attitudes

Similar to beliefs, attitudes have an object, which can be an aspect of the environment like methane gas, or an action such as eating less meat and dairy. However, attitudes differ from beliefs in that they have an affective or emotional, in addition to a cognitive or belief component. Attitudes also have an evaluative component that entails judging an object or action as favorable or unfavorable

(Heberlein 2012). In short, environmental attitudes link beliefs, emotion, and judgment. They can be defined as "*concern* for the environment or caring about environmental issues" (Gifford and Sussman 2012, 65, italics in original), or "a psychological tendency to evaluate the natural environment, and factors affecting its quality, with some degree of favor or disfavor" (Milfont 2012, 268).

Values and beliefs form the foundations for attitudes. An attitude that is based on a greater number of strongly held values and beliefs is harder to change. Further, attitudes that are tied to our identities or based on direct experience are harder to change, whereas attitudes based on fewer, weaker values and that are neither tied to our identities nor based on direct experience are easier to change (Heberlein 2012).

Why Are Values, Beliefs, and Attitudes Important?

> **Even though *individual* attitudes have little to do with *individual* behaviors, they are necessary to support collective actions to change the structure or context of human behavior. That's why it's important to monitor and understand them even if they tell little about what one person . . . would do in specific situations.**
>
> (Heberlein 2012, 68, italics in original)

Definitions of environmental education often assume a linear pathway from knowledge to attitudes to behaviors. However, because the relationships of attitudes, as well as of beliefs and values, to behavior are complex, researchers have turned to other factors (e.g., efficacy, identity, norms) in explaining behaviors. Further, structural factors, such as lack of convenient and well-labeled recycling containers, can negate the relationship of values, beliefs, and attitudes to behaviors (Stern 2000b). Given these caveats, values, beliefs, and attitudes are still important for the following reasons.

- For children, attitudes are associated with environmental behaviors (Eilam and Trop 2012).
- For adults, values and attitudes predict behaviors under certain circumstances, such as when activated, connected to self-concept or identity, and specific to behaviors one wants to change (Bamberg 2003; Steg et al. 2011; Heberlein 2012; Steg, Bolderdijk, et al. 2014).
- Citizens accept and support policies and legislation that are consistent with their values, beliefs, and attitudes (Eilam and Trop 2012; Gifford

and Sussman 2012; Heberlein 2012). Thus, knowing the public's values, beliefs, and attitudes enables governments to formulate policies likely to be accepted by the populace. Similarly, environmental educators can use their understanding of program participants' values, beliefs, and attitudes to plan programs and frame messages.
- Values, beliefs, and attitudes interact with other factors such as norms, identity, and efficacy to influence behaviors.

Author Reflections

I hate litter because I believe that plastics are accumulating in the soil and clogging our waterways with disastrous effects on aquatic life. My negative attitude can also be explained by the fact that litter signals to me disrespectful behaviors—toward the environment and toward other humans.

As I walk to and from work, I pick up litter. Because the Starbucks cups, Cliff Bar wrappers, and Miller Lite beer cans stare me in the face—in other words, litter is ubiquitous and visible—my anti-litter attitudes and biospheric values are constantly activated. But I also experience barriers to picking up litter—it's dirty and gross, and because it's not ordinary for someone to pick up litter on a daily basis, I am concerned that people passing by must think I'm crazy.

Recently, I read about the practice of "plogging" (picking up litter while jogging) and "plalking" (picking up litter while walking) (Dunk, n.d.). Aficionados make the case that these new physical activities are great ways of working out in the outdoors. In addition to holding biospheric values, I value physical fitness, a hedonic value. Knowing that I could improve my fitness not just by jogging and walking, but also by bending down to pick up litter, has reduced the barriers to picking up litter. I now proudly "plog" and "plalk," thinking I am part of a new fitness—and environmental—trend.

How Do Values, Beliefs, and Attitudes Influence Environmental Behaviors?

Values, beliefs, and attitudes influence environmental behaviors and action through multiple pathways, including working alongside self-efficacy, identity, and norms (see chapters 10–12) and by influencing the information we pay attention to (Hines et al. 1986/87; Stern et al. 1999; Hwang et al. 2000; Verplanken and Holland 2002; De Groot and Thøgersen 2013; Van der Werff et al. 2014). When faced with decisions about behaviors that require effort or discomfort, or are costly in terms of time and money, people may let other considerations override their biospheric or altruistic values or environmental attitudes, especially when those values or attitudes are weakly held (Schultz and Oskamp 1996; Steg 2016). The value-norm-belief theory is used to explain the influence of values and beliefs on environmental behaviors (Stern et al. 1999; see also chapter 11), whereas the theory of planned behavior is used to explain how attitudes and beliefs might influence behaviors (Ajzen 1991).

Values and Beliefs

Because values are broad principles, they influence a wide range of behaviors via different mechanisms. Values determine what information we pay attention to, and thus they influence our beliefs. We also draw on our fundamental values when forming attitudes. And biospheric values can contribute to an environmental identity (see chapter 11) as well as activate personal norms of responsibility for environmental problems (see chapter 12). Through influencing beliefs, attitudes, identity, and norms, values indirectly play a role in environmental behaviors. In some instances, values also can directly influence environmental activism (Steg et al. 2011).

The value-belief-norm theory explains how values influence beliefs and norms, which in turn influence behaviors (Stern et al. 1999; we cover norms in depth in chapter 12 but address them here briefly because of their relationship to values and beliefs). This theory stems from work on altruistic values, which suggests that we act on our personal norms when we believe that if we don't act, adverse consequences will befall fellow human beings (Schwartz 1977). Environmental psychologists have extended work on altruistic values to encompass biospheric values, and on social behaviors to encompass environmental behaviors (Stern, Kalof, et al. 1995; Stern 2000b). They posit that values influence three beliefs: ecological worldview (as measured by the NEP), beliefs about consequences of a particular behavior, and beliefs about personal responsibility for environmental degradation. These beliefs in turn determine moral feelings of responsibility to act pro-environmentally—that is, one's personal environmental norms. Finally, personal norms determine our environmental behaviors. In short, people with biospheric values tend to hold environmental worldviews, are aware of the environmental consequences of their behavior, and believe they can reduce these consequences through environmental behaviors; these beliefs lead to a personal environmental norm and eventually to environmental behaviors (figure 7.2; Steg 2016). The value-belief-norm pathway may be particularly important in

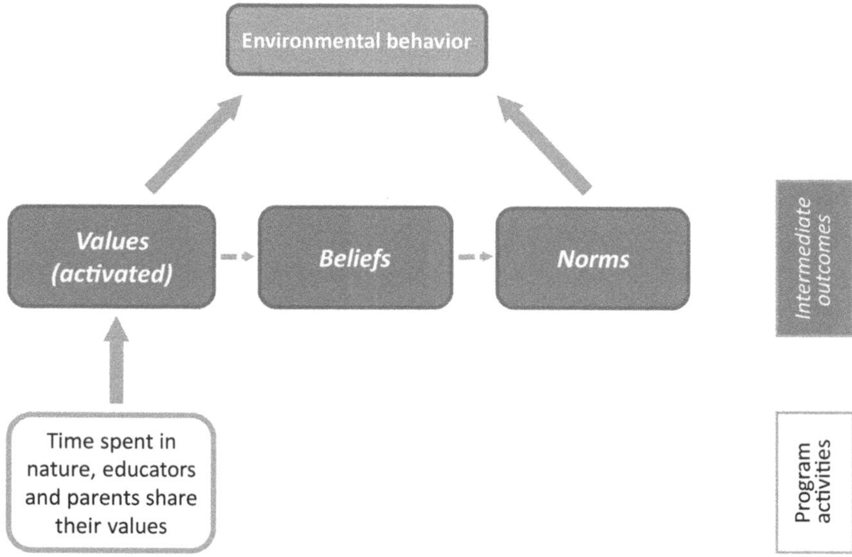

FIGURE 7.2. Values-belief-norms pathway to environmental behaviors. Values, especially when activated, can lead to environmental behaviors. Biospheric values can also influence environmental beliefs, which influence personal norms, which influence environmental behaviors.

explaining low-cost or easy to implement behaviors. For high-cost behaviors, people often believe they cannot make a difference or deny their responsibility for the environmental problem (Stern et al. 1999; Steg and Nordlund 2013).

Biospheric and altruistic values, which reflect multiple life goals or principles, have a stronger association with environmental norms and actions relative to environmental worldview (beliefs), which only captures environmental concerns. In fact, biospheric and altruistic values may directly impact personal environmental norms and environmental activism, without necessarily impacting beliefs (Steg et al. 2011).

Because we all hold multiple and sometimes conflicting values, we prioritize values in decisions about behaviors. One factor that determines which values people prioritize is how central a value is to their identity (Verplanken and Holland 2002; see chapter 10). For example, I may subscribe to environmental and equity values, but environmentalism is more central to my identity, and thus more likely to motivate a behavior like writing an opinion piece for a newspaper. People for whom altruistic values are central to their identity may decide whether or not to support an environmental policy based on its fairness to low-income or minority ethnic groups and whether it promotes equity (Schuitema and Bergstad 2013). Interestingly, when individuals have an opportunity to build an argument in support of their values, they are more likely to act consistently with those values (Maio et al. 2006).

When values are central to our identity or activated, they influence behaviors, as well as beliefs and attitudes. This is because they serve as a filter for what information we pay attention to and even believe (Stern and Dietz 1994; Stern, Kalof, et al. 1995; Thøgersen and Ölander 2006; De Groot and Thøgersen 2013). In fact, one can activate biospheric values by linking them to a person's environmental identity (Verplanken and Holland 2002; Van der Werff et al. 2014). Similar to how a conservative identity can influence what we believe (e.g., about climate change), people with egoistic values may reject information about the environmental consequences of their behaviors. People with biospheric values are not immune to filtering information; they may selectively exaggerate the negative impacts of humans on the environment (Stern and Dietz 1994; Steg 2016).

In short, values influence environmental behaviors through their impacts on our beliefs about environmental issues, about the consequences of our behaviors, and about who is responsible for addressing those issues. These beliefs in turn determine environmental or other norms. Values have the greatest impact on norms and behaviors when they are activated through environmental and other messages or linked to our identity. However, similar to other measures, values do not always lead to environmental behaviors, in part because people have multiple conflicting values that they prioritize in any given situation (Maio et al. 2006; Steg, Bolderdijk, et al. 2014).

Attitudes and Beliefs

Values influence how people form new attitudes when faced with new environmental concerns. This process starts when environmental organizations and other information providers emphasize particular consequences of a new environmental problem (or attitude "object") and thus activate certain values. For example, the Children & Nature Network talks about the health consequences of lack of exposure to nature among city residents, thereby activating altruistic and egocentric values. In contrast, the Nature Conservancy might focus attention on the importance of urban natural areas for migrating birds, thus activating biospheric values. When new issues arise, such as declining urban green space, how groups convey information about the impacts of those issues influences the formation of people's attitudes.

Whereas the value-belief-norm theory emphasizes the importance of values and beliefs, attitudes play a major role in the theory of planned behavior

(Ajzen 1991). This theory posits that the intent to engage in a behavior predicts whether or not someone actually enacts the behavior. Intent, in turn, is determined by attitudes toward the particular behavior, as well as beliefs about how much control we have over a particular behavior and about how others will view the behavior (Ajzen 1991). The theory of planned behavior is more likely to predict behaviors when it uses attitudes toward specific behaviors (Kaiser et al. 2005). For example, a specific attitude about urban nature conservation, in contrast to a general environmental attitude, is more likely to predict our intent to conserve urban natural areas. However, general environmental attitudes do seem to influence beliefs relevant to our intent to adopt environmental behaviors, including beliefs about the consequences of, ability to control, and how others view these behaviors. Thus, general environmental attitudes may indirectly influence our intent to perform environmental behaviors (Bamberg 2003), but keep in mind that what we intend to do is not always what we end up doing.

Several types of attitudes are more closely associated with behaviors. These include attitudes based on direct personal experience with the attitude object, which are likely to be activated in the presence of the object (Maio et al. 2006). For example, my attitudes toward litter are continually activated on my daily walk to work, as I constantly see new litter. This has led me to adopt the behavior called "plalking," or "picking up litter while walking." Ambivalent attitudes, where an individual both positively and negatively assesses an attitude object, have limited influence on behaviors. In contrast, embedded attitudes supported by multiple beliefs, emotions, and experiences, and attitudes formed over time, are more likely to impact behaviors (Maio et al. 2006; Heberlein 2012). In one weeklong environmental education program, participants who entered the program with strong environmental attitudes, presumably developed over time, were likely to carry out environmental actions afterward, whereas those whose environmental attitudes improved during the program demonstrated fewer environmental actions post-program (Ernst et al. 2017).

Some researchers claim that rather than assuming attitudes can lead to behaviors, we should switch our thinking to focus on how behaviors can lead to attitudes (figure 7.3). They point to social marketing and related behavior change campaigns that have been shown to change behaviors in a relatively short time. The researchers also draw on the fact that behaviors can be reinforced by laws and social pressure, whereas this is not generally the case for attitudes. For example, a social marketing campaign in Israel dramatically reduced the number of people picking wildflowers, resulting in recovery of near-extinct plant species. Such behaviors repeated over time may foster environmental attitudes (Eilam and Trop 2012).

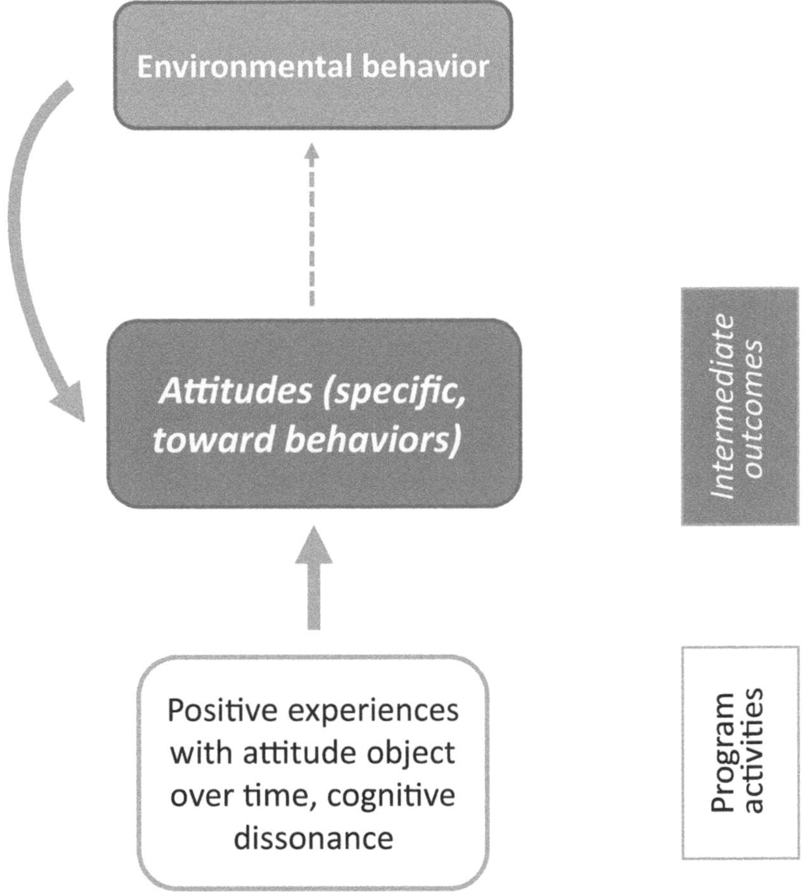

FIGURE 7.3. Specific attitudes and attitudes toward behaviors have a stronger connection to environmental behaviors compared to general attitudes or attitudes toward nature. Attitudes can also be formed through engaging in behaviors and actions.

How Can Environmental Education Foster Values, Beliefs, and Attitudes?

> **Attitudes are like rocks in a rapids. Trying to solve environmental problems by changing attitudes is a little like packing dynamite on a canoe trip and trying to blow up every rock in your way. It's better to read water and avoid collisions with rocks.**
>
> (Heberlein 2012, 10)

By providing opportunities for young people to interact in positive ways with the environment and with adults, environmental education can help children form biospheric and altruistic values, and environmental beliefs and attitudes (Chawla and Cushing 2007; Chawla and Derr 2012; Eilam and Trop 2012; Duarte et al. 2017). Changing adults' more strongly held values, beliefs, and attitudes is not always a realistic goal; however, providing opportunities for adults to engage in environmental behaviors and to repeatedly encounter new information that conflicts with existing beliefs may lead them to reevaluate their values and attitudes over time. The fact that many adults already hold biospheric and altruistic values and positive environmental attitudes opens up opportunities for environmental educators to activate these sentiments to create support for environmental policies and to foster other environmental behaviors (Heberlein 2012; Steg 2016). Thus, becoming familiar with adult participants' values, beliefs, and attitudes enables educators to craft messages and plan programs that resonate with specific audiences. Below we draw primarily on research about changing environmental attitudes, and to a lesser extent on changing values and beliefs.

Developing Environmental Attitudes and Values

Programs that seek to develop new environmental attitudes can provide direct positive experiences with the attitude object—be it an ecosystem, such as a wetland, or an environmental behavior or action, such as restoring a wetland or advocating for its preservation. Importantly, attitudes are formed through a constructivist process, involving social interactions, experiences with the environment, and reflection *over time* (Eilam and Trop 2012; Heberlein 2012). Thus, educators can provide multiple positive experiences, often starting with attitude objects that are somewhat familiar, such as pollinators or gardening, and then moving to more distant objects like advocating for pollinator habitat.

In addition to influencing the formation of attitudes, direct experience may impact children's values and beliefs. However, the opinions of or information provided by parents, the media, teachers, faith-based institutions, and peers—or indirect experiences—generally play a larger role in the formation of values and beliefs (Maio et al. 2006). Thus educators can design hands-on activities specifically around attitude objects they want to change—for example, a consumer behavior—but also should provide information and engage audiences in developing arguments to support related values and beliefs, as well as be aware of parents' and others' influences on attitudes, values, and beliefs.

Because strongly held attitudes are more likely to endure and lead to behaviors, environmental educators need to reinforce attitudes developed during a program. For example, educators can engage children in specific, easy to achieve actions, guide them in reflecting on their actions and the importance of those actions, and provide social support (e.g., through social media groups) after participants return home (Ernst et al. 2017). Because an environmental education program is only one among multiple social influences on the development of environmental attitudes, values, and beliefs, efforts undertaken by schools or environmental education centers can also include family, friends, and community members, who provide support for environmental attitudes beyond a particular program (Eilam and Trop 2012; Duarte et al. 2017). Educators might also leverage youths' and other audiences' altruistic values, by helping audiences see how their environmental behaviors help people as well as the environment.

Cognitive dissonance is a strategy educators can use to foster environmental attitudes in situations where it is easier to change behavior than attitudes. Based on past experiences, people have expectations about things that go together and things that do not. When such expectations are not fulfilled, such as when our behavior seems to contradict our attitudes or values, people experience cognitive dissonance (Festinger 1962). Because we cannot undo a behavior once we have enacted it, youth and adults who engage in an environmental behavior but have anti-environmental attitudes will experience cognitive dissonance. Youth and other audiences may reduce this dissonance by convincing themselves that the actual behavior reflects their true attitudes. In so doing, they may develop new more favorable environmental attitudes (Eilam and Trop 2012).

Working with Existing Values, Beliefs, and Attitudes

> **Internal person dimensions like attitudes, perceptions, and cognitions are difficult to define objectively and change directly. So stop trying! . . . Instead, look for external factors influencing behavior independent of individual feelings, preferences, and perceptions. When you empower people to analyze behavior from a systems perspective and to implement interventions to improve behavior, you will indirectly improve their attitude, commitment and internal motivation.**
>
> (Geller 2002, 528)

Educators working with adults or youth who already have environmental values and attitudes can support and activate these values and attitudes. When working with people who do not hold environmental values and attitudes, educators can "navigate" existing values and attitudes to achieve environmental outcomes.

Because individuals with strong attitudes are more likely to engage in behaviors, environmental education programs whose primary goal is to improve environmental quality may want to target youth or adults who already have strong environmental attitudes (Ernst et al. 2017). However, this strategy may pit environmental education's biospheric values and commitment to environmental conservation against its altruistic values and commitment to equity and diversity, or to providing programs for youth with fewer opportunities (Delia and Krasny 2018; NAAEE, n.d.). Further, environmental educators often see their job as improving environmental attitudes, rather than providing behavior options for audiences who already support environmental causes. Still, building the capacity of youth and other audiences who already hold environmental values and attitudes to take action can be a powerful strategy to achieve environmental outcomes.

When faced with decisions about behaviors that require effort, discomfort, or are costly in terms of time and money, people holding biospheric values may let other considerations override. Here environmental educators can support action by increasing the benefits and reducing the costs of acting on biospheric values. Strategies include creating incentives and rewards for environmental behaviors, focusing on less costly or more pleasurable environmental activities, providing feedback on the costs and benefits of environmental behaviors, and creating environmentally friendly defaults, such as providing water refill stations and not providing single-use plastic bottles (Schultz and Oskamp 1996; Sunstein and Reisch 2014; Steg 2016; see also chapter 12).

A strong environmental work culture also can encourage workers to express as well as strengthen existing biospheric values (Ruepert et al. 2017). Environmental education programs can demonstrate this culture through mission statements, leaders making hard decisions when money-making goals come up against environmental goals, installing energy-saving or water-saving technology like low-flush or composting toilets, serving vegetarian or vegan meals and avoiding onetime plastic dishes and cutlery, and sponsoring corporate responsibility or volunteer days. In short, environmental education programs can model environmental practices at their nature center or other educational setting, and provide activities that appeal to people holding a range of values (e.g., volunteering in a community garden can appeal to those holding altruistic values).

Activating and strengthening existing biospheric and altruistic values can also influence behaviors. Publicly endorsing biospheric and altruistic values, reading a story or playing a game that focuses on the environment, and helping audiences to develop and articulate arguments that support positive values and to link these values to their identity are all strategies to activate values (Verplanken and Holland 2002; Maio et al. 2006). Further, engaging audiences in action that is consistent with biospheric and altruistic values can instill feelings of being a moral person, thus activating those values and reinforcing personal norms (see chapter 12). Relatedly, environmental educators can consider which values are activated in a particular setting or activity—e.g., hedonic or biospheric—and focus audiences' attention on the related personal or social norm consequences of behaviors (Steg et al. 2011; Steg and de Groot 2012; Steg, Perlaviciute, et al. 2014; Steg, Bolderdijk, et al. 2014). Finally, educators can help participants identify barriers to action, including prevailing attitudes, norms, and beliefs; conflicting values and attitudes and the need to prioritize them; and structural or technological barriers such as lack of low-cost energy or clean water alternatives to single-use plastic water bottles (Stern et al. 1999; Verplanken and Holland 2002; Heberlein 2012; Ernst et al. 2017).

Focusing on positive youth and community development is a means of finding common ground with educators driven by altruistic values and with parents driven by egoistic values expressed by wanting to see their children succeed (Stern and Dietz 1994). Programs that build both environmental literacy and youth assets simultaneously engage youth in community gardening, tree planting, water quality, and other restoration efforts and in addressing environmental injustice (Schusler and Krasny 2008; Schusler et al. 2009; Schusler and Krasny 2010; Schusler 2014; Schusler and Krasny 2014; Aguilar et al. 2015; DuBois et al. 2017; Russ and Krasny 2017; Delia and Krasny 2018; see also chapter 14).

Finally, educators can "navigate" existing anti-environmental attitudes and beliefs to engage audiences in environmental behaviors (Heberlein 2012). Environmental educator Anne Armstrong worked in coastal Virginia with residents who did not believe in climate change and had negative attitudes toward climate scientists, but who shared the altruistic value of wanting to help their community. Rather than try to change her audiences' beliefs and attitudes about climate change, Anne gave them an opportunity to participate in oyster reef restoration activities to help stabilize the local eroding shoreline. In this way, she drew upon their concern that their community was suffering from shoreline erosion due to sea level rise but did not activate contentious beliefs about what was causing the sea level rise, which likely would have turned residents off to any environmental engagement. Environmental sociologist Thomas Heberlein has likened working with existing attitudes to canoeing down rapids—the best approach is go with rather than fight the flow, but be on the lookout for eddies or counter-currents that provide opportunities (Heberlein 2012). In short, the key to changing behaviors is to choose behaviors it might be possible to change and then to engage people in light of what you know about their attitudes and beliefs, but without necessarily trying to change those attitudes and beliefs (Heberlein 2012).

Assessing Values, Beliefs, and Attitudes

Values, beliefs, and attitudes are generally measured using Likert scale survey questions (see appendix). Values questions ask respondents to indicate where they fall on a continuum between being opposed to a particular value and the value being an extremely important life guiding principle (Stern, Kalof, et al. 1995).

The New Environmental Paradigm (Dunlap and Van Liere 1978) or revised New Ecological Paradigm (Dunlap et al. 2000) is the most commonly used measure of beliefs. (Some researchers do not distinguish between beliefs and attitudes, and use this scale as a measure of attitudes.) The scale includes items about earth's ecological limits (e.g., we are approaching the limit of the number of people the earth can support); balance of nature (e.g., the balance of nature is strong enough to cope with the impacts of modern industrial nations); human domination (e.g., humans were meant to rule over the rest of nature); and ecological catastrophe (e.g., the so-called "ecological crisis" facing humankind has been greatly exaggerated). However, the scale is outdated, given climate change and more other recent environmental issues (Bernstein 2017) and is generally not closely associated with environmental behaviors (Steg et al. 2011); thus I do not recommend using the New Environmental Paradigm to predict behaviors.

Measuring attitudes entails asking about the extent to which one feels favorable or unfavorable toward the environment or an environmental action (Maio et al. 2006). Published scales often measure general environmental attitudes, which are important in predicting public acceptance of environmental policies. If you want to use attitudes to predict the likelihood of audiences engaging in particular environmental behaviors, you can adapt questions to reflect the places, issues, and possible behaviors specific to your program.

The Environmental Attitude Inventory measures twelve constructs, including some that focus solely on cognitive aspects and might be considered beliefs (Milfont and Duckitt 2010). The questions that include affective and judgment in addition to cognitive components, and thus more closely reflect attitudes, include those about enjoyment of nature (e.g., I think spending time in nature is boring); environmental movement activism (e.g., I would like to join and actively participate in an environmentalist group); altering nature (I'd prefer a garden that is wild and natural to a well groomed and ordered one); and ecocentric concern (e.g., It makes me sad to see forests cleared for agriculture).

8

NATURE CONNECTEDNESS

> **Can people experience a personal relationship with the environment analogous to how they experience a relationship with another human being?**
>
> (Davis et al. 2009, 173)

Highlights

- Nature connectedness is a feeling of being connected and belonging to the natural community.
- Nature connectedness fosters environmental behaviors through its association with feelings of belonging to the community of nature, of nature being part of our identity, and of happiness.
- Environmental education can foster nature connectedness among children through providing long-term, repeated, sensory experiences in nature, often with family members.

Nature connectedness captures the emotional component of human-nature interactions. Because feelings of unity or communion with nature can lead to empathy for other organisms, nature connectedness is often a precursor of environmental concerns and behaviors (Dutcher et al. 2007).

What Is Nature Connectedness?

> **If people feel connected to nature, then they will be less likely to harm it, for harming it would in essence be harming their very self.**
>
> (Mayer and Frantz 2004, 512)

Nature connectedness can be defined as a feeling of being connected and belonging to the natural community (Mayer and Frantz 2004). Feelings of being connected to nature are rooted in biophilia, or humans' innate "love" of nature (Wilson 1984; Kellert and Wilson 1993), and in Aldo Leopold's notions of the land as a community of "soils, waters, plants, and animals" of which humans are "plain member and citizen" (Leopold 1949).

Whereas we use the term "nature connectedness" in this chapter, a number of related concepts capture humans' ties to nature (table 8.1). Some of these constructs focus solely on the emotional or affective, whereas others also include cognitive components. An example of the former is emotional affinity toward nature, which is distinguished from its cognitive counterpart, interest in nature. Environmental psychologists explain this difference: "One can have scientific interest in nature issues without feeling any emotional affinity. *Interest* motivates gathering knowledge to explain and understand phenomena. *Emotional affinity* is motivating contact and sensual experiences" (Kals et al. 1999, 182, emphasis in original).

Nature relatedness encompasses feelings toward nature as well as cognitive worldviews of nature and experiences in nature. It is defined as "one's appreciation for and understanding of our interconnectedness with all other living things on the earth" (Nisbet et al. 2009, 718). These and related constructs, including inclusion of nature in self (Schultz 2001; Schultz et al. 2004) and commitment to the natural environment, expand on earlier notions of how humans are connected to and depend on each other, to encompass human-nature interconnectedness and dependence (Davis et al. 2009).

Nature connectedness, emotional affinity toward nature, and nature relatedness all focus on human-nature interactions, and in this way can be distinguished from related constructs discussed in this book. For example, sense of place, which includes place attachment and place meaning, captures the social and cultural aspects of our surroundings in addition to natural elements (Stedman 2002; Kudryavtsev et al. 2011; see chapter 9). The cognitive construct systems thinking entails realizing that humans are part of natural systems but lacks the affective component of nature connectedness (Thibodeau et al. 2016; Otto and Pensini 2017; see chapter 6).

Nature connectedness and related concepts are often tied to collective, environmental, or ecological identity (Schultz and Tabanico 2007; Nisbet et al. 2009; Gosling and Williams 2010; Brügger et al. 2011; Tam 2013; see chapter 11). For example, the construct inclusion of nature in self attempts to capture the "extent to which an individual includes nature within his or her cognitive representation of self" (Schultz and Tabanico 2007, 1221), with our representations of self being closely tied to our identities.

TABLE 8.1 Nature connectedness and related constructs

CONCEPT	DEFINITION	COMPONENTS
Connectedness to nature (Mayer and Frantz 2004)	Feeling of being connected and belonging to the natural community	Affective
Emotional affinity toward nature (Kals et al. 1999)	Emotional inclinations toward nature such as love for nature, feelings of freedom and safety in nature, and feeling of oneness with nature	Affective
Nature relatedness (Nisbet et al. 2009)	Appreciation for and understanding of our interconnectedness with all other living things on earth	Affective, cognitive, and experiential
Inclusion of nature in self (Schultz 2001; Schultz et al. 2004)	Extent to which one thinks of oneself as including aspects of nature	Cognitive
Connectivity with nature (Dutcher et al. 2007)	Seeing environment as part of self and self as part of environment, reflects empathy due to unity/ communion between self and nature	Affective
Commitment to the natural environment (Davis et al. 2009)	Psychological attachment to and long-term orientation toward the natural world	Affective
Environmental identity (Clayton 2003)	Belief that the environment is important to us and an important part of who we are	Multidimensional

Why Is Nature Connectedness Important?

> **When we see land as a community to which we belong, we may begin to use it with love and respect.**
>
> (Leopold 1949, viii)

Environmental psychologists have found that nature connectedness predicts environmental behaviors across multiple audiences and contexts.

- Nature connectedness and related constructs (table 8.1) are strong predictors of environmental and nature-protective behaviors among children, college students, and adults (Kals et al. 1999; Nisbet et al. 2009; Cheng and Monroe 2012; Tam 2013; Frantz and Mayer 2014). They are more strongly associated with environmental behaviors relative to other constructs, including the New Environmental Paradigm (Dunlap and Van Liere 1978, 2008), biospheric values, and environmental knowledge or systems thinking (Finger 1994; Dutcher et al. 2007; Frantz and Mayer 2014; Davis and Stroink 2016; Otto and Pensini 2017).

- Nature connectedness is closely linked to environmental identity (see chapter 11), which exerts strong influences on environmental behaviors (Tam 2013).
- Nature connectedness has multiple health benefits, including being associated with feelings of happiness (Nisbet and Zelenski 2011; Zelenski and Nisbet 2012; Capaldi et al. 2014, see chapter 15). The "happy path to sustainability" (Nisbet and Zelenski 2011) enables environmental educators to "put a more positive spin on ecological behavior than the doom and gloom messages that warn the public to change or die. . . . A positive framing may in the long run provide a more effective means of promoting environmentally friendly behavior" (Mayer and Frantz 2004, 512).

Author Reflections

When I was a child, my parents took my brothers and me hiking along the Billy Goat Trail near Washington, DC. Our family, including my grandfather, hiked and canoed during "summer camp" sponsored by the Appalachian Mountain Club. There an older lady taught me the names of wildflowers. And I remember developing a deep appreciation for nature the summer I hiked from hut to hut with my adopted Austrian family in the Alps. In short, plenty of adults shared their love of nature with me when I was a child.

My college graduation present was a monthlong mountaineering expedition in the Glacier Peak Wilderness of Washington State. For the next three summers, I led similar expeditions for the National Outdoor Leadership School—scaling glaciers, crossing rain-fed torrents, and wandering through alpine meadows (figure 8.1). My connection to nature was profound.

Then I moved to Ithaca, New York. Although many appreciate the gorges, forests, and hilly agricultural landscape, I was not so entranced with this more "cultivated" nature of rural New York State. And yet, over the years—through hiking and canoeing with my own family, and hopefully instilling a feeling of nature connectedness in my children—I have become increasingly connected to this local nature. As I have grown older, my connection has become more individual—emerging through my early morning walks and weekend runs listening to the sound of waterfalls, feeling the calm of a dark woods, and dodging the occasional skunk rooting for worms after a rainfall.

FIGURE 8.1. Author reveling in her connection to nature. Photo by Alex Russ.

How Does Nature Connectedness Foster Environmental Behaviors and Collective Actions?

> **All ethics so far evolved rest upon a single premise: that the individual is a member of a community of interdependent parts. . . . The land ethic simply enlarges the boundaries of the community to include**

> **soils, waters, plants, and animals, or collectively: the land. . . . A land ethic changes the role of *Homo sapiens* from conqueror of the land-community to plain member and citizen of it. It implies respect for his fellow-members, and also respect for the community as such.**
> (Leopold 1949, 203–204)

Nature connectedness can influence environmental behaviors through two pathways, both of which draw on affect and relationships. The "we-ness" pathway expands our feelings of connectedness to other humans to encompass connectedness to all other beings (Frantz and Mayer 2014). The "happiness" pathway is based on the health and well-being outcomes of spending time in nature (Zelenski and Nisbet 2012).

According to the "we-ness" pathway (figure 8.2), as we become closer to other individuals and they become part of how we define ourselves, we demonstrate greater empathy and willingness to help (Cialdini et al. 1997). Similarly, as we

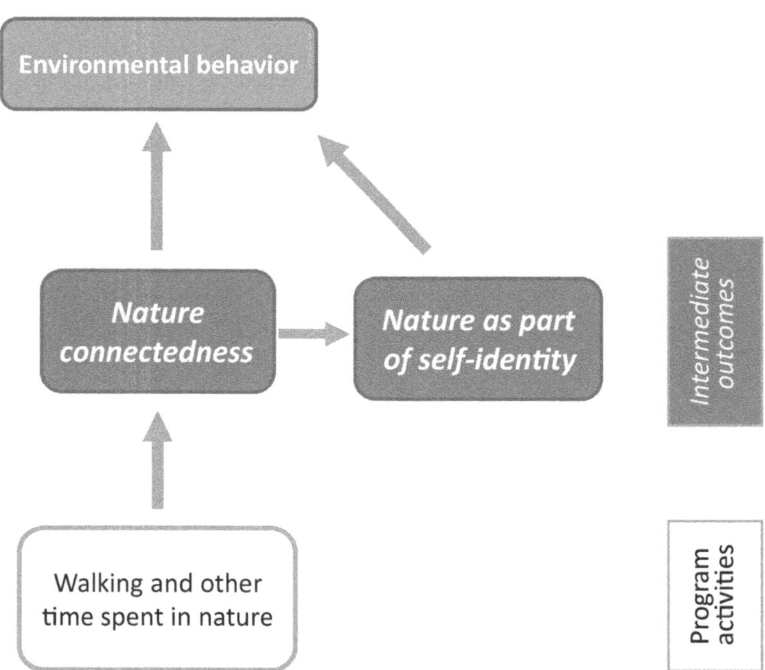

FIGURE 8.2. The "we-ness" pathway to sustainability. People who spend time in nature develop a connection with nature, and nature becomes part of their identity, which leads to environmental behaviors (Frantz and Mayer 2014).

become closer to nature—and nature becomes part of our identity or "self"—we feel more empathy toward, concern about, and willingness to help nature (Schultz 2001; Schultz and Tabanico 2007; Gosling and Williams 2010; Frantz and Mayer 2014). Feelings of belonging to the community of nature, or of nature being part of our identity, play a role because harming nature feels akin to harming ourselves (Beery and Wolf-Watz 2014). The we-ness pathway draws from the work of Aldo Leopold, who rather than drawing a sharp line between humans and nature, spoke about humans as citizens of the "land community" (Leopold 1949). In Leopold's view, nature is no longer some "other," for whom we care little and thus can justify taking actions against (similar to how conceiving our presumed enemies as "other" enables violent behaviors). If we feel as if nature is part of our community, or that we are part of the land community, we accept our role in that community, and acting on its behalf becomes acting on our own behalf (Goralnik and Nelson 2011).

The "happiness" pathway builds on social sciences research suggesting that connectedness with family and friends is associated with happiness (figure 8.3). It turns

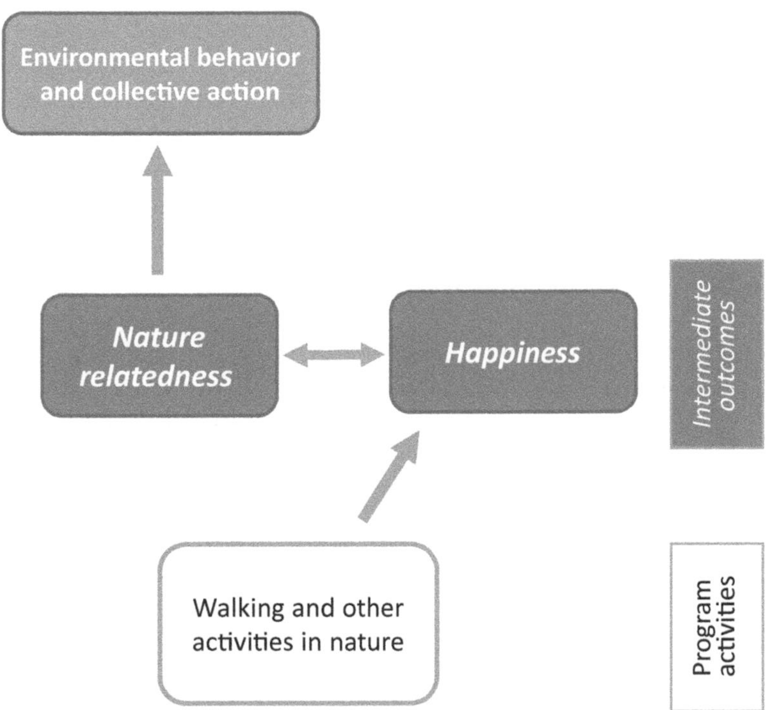

FIGURE 8.3. The "happiness" pathway to sustainability. Time spent in nature spurs happiness and nature relatedness, which makes people more inclined to protect nature (Zelenski and Nisbet 2012).

out that people who feel a sense of nature relatedness also feel happier. In fact, the happiness associated with a sense of nature relatedness might even offset some of the distress that accompanies an awareness of environmental destruction. The link between nature relatedness and happiness also would explain how hedonic and egoistic values—that is, realizing that nature is important to our well-being—could lead to environmental behaviors (Nisbet and Zelenski 2011; Zelenski and Nisbet 2012).

How Can Environmental Education Foster Nature Connectedness?

Environmental educators should focus on three factors in planning programs to connect people with nature: the amount of time program participants will be able to spend in nature, how that time is spent, and participants' age. Longer-term, repeated experiences with close family members are ideal; shorter walks or activities in nature and spending time with friends and teachers have a more limited impact. Sensory activities, such as walking in streams and sitting quietly in the woods, as opposed to purely cognitive activities that focus on learning about nature, have been shown to increase nature connectedness. Finally, younger children up to about age eleven are more likely to develop feelings of nature connectedness relative to older children and adults (Chawla and Cushing 2007; Ernst and Theimer 2011; Cheng and Monroe 2012; Liefländer et al. 2013).

Adults spending time in nature with children can communicate and transfer positive emotions related to the natural environment (Kals et al. 1999; Chawla and Cushing 2007; Nisbet and Zelenski 2011; Otto and Pensini 2017). Further, educators might want to consider using language that suggests participants are protecting themselves, or their "home," rather than protecting "other" places apart from our real world. For example, educators can lead students who have seen a butterfly in a discussion of feeling a sense of kinship or community, possibly even likening it to kinship we might feel for a friend or family member, as opposed to emphasizing the "otherness" of the butterfly or nature more broadly (Goralnik and Nelson 2011). Educators and parents can also talk about ethics and responsibility related to taking care of nature (Kals et al. 1999).

How Can We Assess Nature Connectedness?

There are as many scales to measure nature connectedness as there are nature-connectedness-related constructs (table 8.1; appendix). Not surprisingly, these scales measure similar concepts (Brügger et al. 2011; Tam 2013). Here we focus on connectedness to nature (Mayer and Frantz 2004), inclusion of self in nature (Schultz 2001; Schultz et al. 2004), and nature relatedness (Nisbet et al. 2009), including scales that have been adapted for children (Ernst and Theimer 2011; Cheng and Monroe 2012).

The Connectedness to Nature scale is based on Leopold's vision of a sense of kinship with and belongingness to nature. It includes propositions such as "I often feel a sense of oneness with the natural world around me"; "I think of the natural world as a community to which I belong"; and "I recognize and appreciate the intelligence of other living organisms" (Mayer and Frantz 2004).

A connection-to-nature scale designed specifically for children includes statements that reflect children's enjoyment of nature (e.g., "I like to hear different sounds in nature"); empathy for its creatures ("I like to see wild animals living in a clean environment"); sense of oneness ("Humans are part of the natural world"); and sense of responsibility for nature ("My actions will make the natural world different") (Cheng and Monroe 2012). Another approach to assess nature connectedness in children uses descriptions of two young people, one of whom is connected to nature and the other who isn't, and asks children to first choose which of the two people they are most like, and then to rate how much they are like that person. For example, a statement might read, "Some kids like to spend their weekends outside walking in parks, but other kids like to spend their weekend inside" (Musser and Malkus 1994; Ernst and Theimer 2011).

The measure for "inclusion of nature in self" focuses specifically on the degree to which humans include nature in how they represent themselves. Respondents view seven diagrams, each consisting of a circle labeled "self" and a circle labeled "nature." The diagrams vary from complete separation of the two circles to complete overlap, and respondents choose which diagram along the continuum best represents their relationship with the natural world (Schultz 2001). Although diagrams may be easy to use with children, the overlap of nature and self is an abstract concept and has a weaker relationship with environmental behaviors compared to other nature connectedness measures (Brügger et al. 2011).

The "nature relatedness" survey assesses one's personal connection to nature, nature-related worldview, sense of agency concerning human actions and their impacts on nature, and physical familiarity or comfort with being in nature (Nisbet et al. 2009). Questions might need to be adapted for a particular audience; for example, some children or adults might not be familiar with or have access to wilderness areas, yet might enjoy spending time in city parks (Nisbet and Zelenski 2013).

We also can assess nature connectedness by measuring nature-related behaviors (Brügger et al. 2011). For example, respondents might indicate how frequently (from never to very often) they do the following: take time to consciously smell flowers; consciously watch or listen to birds; collect objects from nature such as stones, leaves, or insects; take care of plants at home or school; or take walks regardless of the weather. Respondents can also state their level of agreement with statements that reflect appreciation of nature, such as: the croaking of frogs is comforting; listening to the sounds of nature makes me relax; I enjoy gardening; or my favorite place is in nature.

9

SENSE OF PLACE

Highlights

- Sense of place reflects how people perceive and feel about places, including meanings they attribute to places and how strongly they are attached to places.
- People who ascribe ecological meanings to a place are more likely to engage in stewardship and other behaviors to protect a place against development.
- People with strong place attachments act to protect places threatened by development, and to restore places that have been damaged by economic and environmental decline and disaster.
- Environmental educators can foster ecological place meanings by providing opportunities for youth to spend unstructured time in nature, and can foster sense of place more broadly through providing recreational, stewardship, citizen science, and action research experiences in nearby neighborhoods.

The notion of sense of place resonates across cultures. When my former PhD student Alex Kudryavtsev was conducting research in the Bronx, New York City, he listened to educators describe their goals for the youth in their programs. When he told them that their goals sounded like instilling a "sense of place" among youth growing up in this low-income, ethnically diverse neighborhood, the educators were excited to find a term that described their aspirations. Half a

FIGURE 9.1. "I love this river and have proudly served it for over 20 yrs. Sooo blessed to still be in the game. #attitudeofgratitude—at Anacostia River." Akiima Price, by her beloved Anacostia River, November 2018. Photo by Michael Wood.

world away, when I lectured about sense of place to Chinese university students in Beijing, they immediately latched on to the concept—and how it applied to where they grew up and the large city in which they were now living. One student even published an essay on how moving away had changed the way she viewed the place where she was from (Yu 2018b). And my colleague and friend Akiima Price from Washington, DC, has developed a strong place attachment to the river that she grew up near and has helped steward for twenty years (figure 9.1).

What Is Sense of Place?

Sense of place refers to the meanings and emotions we associate with a particular place. It includes both a cognitive (place meaning) and affective (place attachment) component (table 9.1). Place meanings and attachments encompass the physical and biological as well as the social and cultural aspects of a particular place, such as a park, neighborhood, city, or region (Tuan 1974; Stedman 2002; Manzo and Perkins 2006; Kudryavtsev et al. 2011; Ardoin 2014; Rickard and Stedman 2015; Masterson et al. 2017; Larson et al. 2018).

TABLE 9.1 Sense of place and its components (Jorgensen and Stedman 2001; Stedman 2002)

COMPONENT	DEFINITION	EXAMPLES
Sense of place	Meanings and emotions individuals or groups associate with a particular place	Seeing my town as a place where people are friendly and feeling as if it's a place where I can fulfill my life goals
Place meaning	Descriptive or symbolic representations of a place	My city is a place with ample green space, my city is a place that welcomes all
Place attachment	A bond formed with a place based on cognitions (or meanings) and feelings	Seeing my city as dynamic and fun, which reflects who I am and what I like to do
Place identity	Individual or group identity that relates to physical and symbolic attributes of a place	Identifying as being part of an oceanside community
Place dependence	Potential of a place to satisfy needs or help achieve goals	Feeling that one's oceanside community enables one to satisfy one's needs for a calm place to walk

Place meanings are descriptive or symbolic representations of a place. For example, you might describe your neighborhood as a place where you can see wildlife, enjoy outdoor recreation, or spend time with friends. Or you might see where you live as a "verdant city," that is, a symbol of how nature can be integrated into the built environment, or a "welcoming city," symbolic of how a place can be hospitable to immigrants and people of multiple ethnicities (Stedman 2002; Kudryavtsev 2013; Masterson et al. 2017; Larson et al. 2018). Ecological place meanings are meanings attributed to wildlife, plants, and other natural phenomena, and to nature-based activities such as building forts with natural objects, gardening, or canoeing (Kudryavtsev et al. 2012). Depending on the particular meanings people hold, they react differently to threats to a valued place. For example, in a study of people owning vacation homes around a lake in northern Wisconsin, some homeowners saw the lake as a place of natural beauty, whereas others described it as a place to spend time with family and friends. Only the homeowners with ecological place meanings were willing to take action to protect the lake from development (Jorgensen and Stedman 2001).

While recognizing that humans associate both positive and negative emotions with places, place attachment generally refers to positive bonds people form with a place (Altman and Low 1994; Masterson et al. 2017). Place attachment also is described as a feeling about the value and significance of various place meanings (Larson et al. 2018). Children are attached to a place when they "show happiness at being in it and regret or distress at leaving it, and when they value it not only for the satisfaction of physical needs but for its own intrinsic qualities" (Chawla 1994, 64). If I see my neighborhood as a place where I can take pleasant walks in nature, and I enjoy walks in nature, I am likely to develop an attachment to my neighborhood. In short, place attachment describes an attraction to a place, while place meaning provides a foundation for that attraction (Stedman 2002).

In addition to place meaning and place attachment, place identity is often considered a component of sense of place (Jorgensen and Stedman 2001, 2006). Place identity refers to the ways in which physical and symbolic attributes of certain locations contribute to an individual's sense of self or identity (Proshansky et al. 1983). Humans have both individual and social place identities (Uzzell et al. 2002; Devine-Wright 2009; see also chapter 11). In one study, residents who valued and identified with the ecological features of their coastal town resisted the installation of offshore wind turbines, whose presence would have disrupted their place attachment and threatened their place identities (Devine-Wright and Howes 2010).

Another component of sense of place, place dependence, refers to the potential of a place to satisfy people's needs, in particular by providing settings for outdoor recreation. Thus we may depend on a particular park as a place where we walk our dog (Jorgensen and Stedman 2001; Vaske and Kobrin 2001; Farnum et al. 2005). Some researchers consider place identity and place dependence as part of place attachment, while others consider these constructs as more broadly related to sense of place (Altman and Low 1994; Vaske and Kobrin 2001; Ardoin 2006; Kudryavtsev et al. 2011).

People's sense of place is often tied to their sense of community, which is defined as feelings among members of a group that they belong to the group, that they are important to each other, and that their needs will be met through the group (McMillan and Chavis 1986). While sense of place captures meanings and attachment to the physical as well as social aspects of place, sense of community focuses on bonds among people. Often sense of community, and related attributes such as social cohesion and social capital, are found in places where people also have strong place attachment (Pretty et al. 2003; Amsden et al. 2010; Lewicka 2011). However, online communities are increasingly becoming sources of identity, connections, and meaning, and also may foster a sense of community and mobilize both online and offline (or place-based) collective action (Ballew et al. 2015).

As more people are displaced due to conflict, economic stress, or disaster, their sense of place and sense of community can be disrupted (Rivlin 1982; Ardoin 2006). This is captured by the comments made by a resident of a West Virginia town after a devastating flood that displaced many residents: "I have a new home right now, and I would say that it is a much nicer home than what I had before. But it is a house, it is not a home. Before I had a home" (Erikson 1976, 175). Such comments suggest the importance of sense of place as a component of loss, in light of more frequent and devastating disasters.

Author Reflections

I have lived in several places, some of which hold negative place meanings, while others hold positive place meanings. My place meanings in turn influence my attachment to different places. I grew up in the Washington, DC, suburbs, to which I associate several negative meanings, such as lack of feeling of community, car dependent, traffic congestion, and sultry summers. These negative meanings may be part of the reason I don't feel particularly attached to this place. Fortunately for me, I associate positive meanings with where I currently live—Ithaca, New York—including ecological place meanings associated with our scenic gorges, lakes, hawks, ospreys, and walkable neighborhoods, and cultural features such as opportunities for learning through Cornell University. Just this morning on my walk to work I saw a raccoon, numerous rabbits, and a screech owl peering from a nest box next to Beebe Lake. I even smelled a skunk, one of my favorite animals! And as I write from my office, I hear the ravenous fledging red-tailed hawks on the neighboring building screaming at their parents for food. Walking and running through Ithaca's natural areas are part of my identity, and I depend on these places for my recreation and well-being.

Why Is Sense of Place Important?

> **Processes of collective action work better when emotional ties to places and their inhabitants are cultivated.**
>
> (Manzo and Perkins 2006, 347)

People are motivated to protect places that hold personal and valued meanings, places to which they are attached, and places which form part of their identity (Stedman 2002; Cheng et al. 2003; Devine-Wright 2009; Ardoin 2014).

- People are likely to steward a place if they feel both a strong place attachment based on the place's natural features and attribute to that place ecological place meanings (Lewicka 2011; Kudryavtsev et al. 2012; Larson et al. 2018).
- People holding conflicting place meanings or who depend on a place for different reasons—e.g., natural area as a place for motorized or nonmotorized recreation—may draw on their common place identity, alongside trust and social ties, to develop common goals for managing a particular place (Cheng et al. 2003; Payton et al. 2005). For example, two groups of residents in rural Alberta, Canada, one that used the Oldman River watershed for off-road vehicle use and the other for hiking, were both attached to their shared place and worked together to restore the watershed (OWC 2017).
- Meanings attached to an iconic species, for example oysters in New York City, can spur behaviors to protect that species and even rebuild the "places" that provide a home for valued species (e.g., oyster reefs) (Krasny et al. 2014).
- Place attachment to natural areas and to one's neighborhood is associated with general environmental behaviors such as reducing energy or recycling batteries (Vaske and Kobrin 2001; Halpenny 2010; Scannell and Gifford 2010; Rioux 2011).
- Whereas place attachment based on social connections may not predict environmental conservation behaviors (Rivlin 1982; Scannell and Gifford 2010), it can be leveraged as a tool for neighborhood revitalization (Brown, Perkins, et al. 2003). Neighborhood revitalization projects that include community gardening, litter cleanups, and other civic ecology practices provide settings for environmental education (Krasny and Tidball 2009a, b; Krasny, Lundholm, et al. 2013).
- Place attachment has psychological benefits, including positive memories and emotions and feelings of belonging, relaxation, and comfort (Lewicka 2011; Scannell and Gifford 2017).

How Does Sense of Place Influence Environmental Behaviors and Collective Action?

> **Social and political behaviors and place meanings are not discernable by looking solely at biophysical attributes or individual inhabitants of the place; they emerge as result of the interaction between biophysical attributes and social and political processes.**
>
> (Cheng et al. 2003, 99)

When residents' place meanings, attachment, and identity are threatened by gradual decline of a neighborhood, they may be spurred to take place-protective collective action. This can happen when employers go elsewhere, houses are abandoned and boarded up, and graffiti stares out from crumbling concrete walls. Sudden disasters, such as floods, can also spur action to rebuild. Alternatively, residents facing decline and disaster and other threats to their sense of place may become despondent and neither initiate nor respond to calls for action.

Whether or not people engage in place-protective behaviors in response to decline and disaster depends not only on their sense of place, but also on the level of existing community ties (e.g., sense of community, social cohesion, or social capital; figure 9.2; see also chapter 13). Thus, sense of place can spur collective actions through a combination of threats to sense of place and social cohesion (Uzzell et al. 2002; Brown et al. 2003; Payton et al. 2005; Manzo and Perkins 2006; Devine-Wright 2009; Lewicka 2011). Those who are more attached to their neighborhoods tend to interact more with neighbors, suggesting that place attachment and community ties are created simultaneously (Manzo and Perkins 2006). Note that in addition to drawing on community ties, conservation action reinforces existing and even builds new and stronger community ties. This occurs through people communicating about the impact of proposed changes on the environment and the people who live there (Devine-Wright 2009) and through working together to restore valued places (Krasny and Tidball 2015). However, in some cases disruption leads to a weakening of both social networks and place attachment, especially if feelings of loss are ignored (Manzo and Perkins 2006).

Place-protective behaviors also emerge in neighborhoods and rural areas threatened by industrial, energy, and other types of development. NIMBY, or "not in my back yard," is a term used to refer to people protesting when an industrial or other facility is proposed near where they live. Similar to other place-protective behaviors, local opposition is sparked when change disrupts emotional attachments to place and place identity (Devine-Wright 2009). Often it is a combination of several threats that spurs people to action.

In short, place attachment and community ties are inextricably linked in part because place attachment depends not only on the physical environment but also on our experiences with others in the community. Place-protective action often occurs when place attachment is disrupted, especially in neighborhoods with strong social cohesion and where feelings of loss are addressed (Manzo and Perkins 2006; Devine-Wright 2009; Lewicka 2011). Place identity, place attachment, and social cohesion work hand in hand to foster place-specific environmental behaviors and collective action (Uzzell et al. 2002).

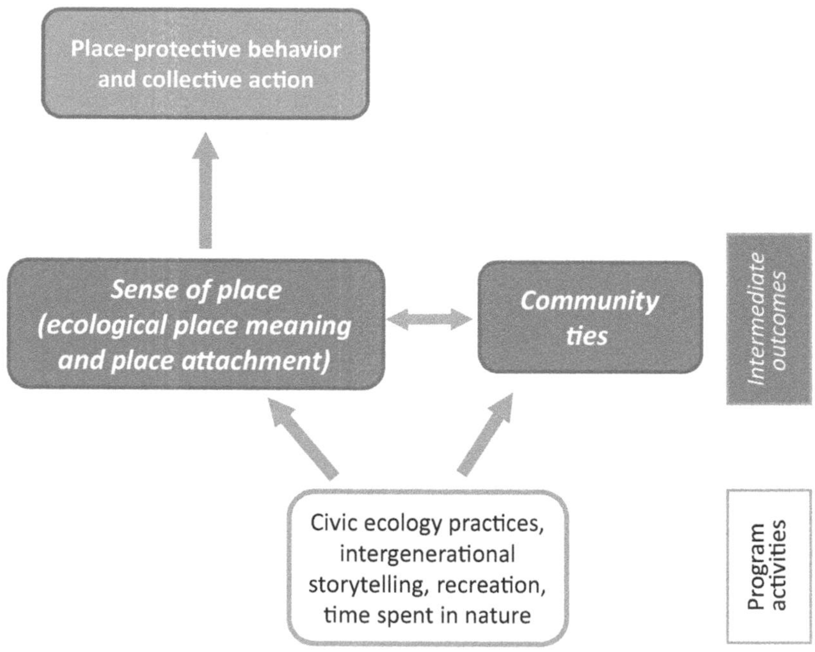

FIGURE 9.2. Sense of place works hand in hand with community ties to spur place-protective collective action, especially when residents experience a threat or disruption to their community.

How Can Environmental Education Nurture Sense of Place?

> **In today's increasingly transient world, a rooted, ancestral connection to place is becoming increasingly rare. Therefore, place-based education programs may be most effective when they recognize the diversity of place attachments that exist and cumulate from a range of relationships with the landscape. . . . Place-based education should strive to reach a range of community members through building on individual, unique perspectives, rather than privileging only a rooted sense of place.**
>
> (Ardoin 2006, 120)

People develop place meanings and place attachment through positive interactions with a place and with the people in that place. Participants in an environmental education program will already have developed place meanings through their past experiences. In everyday activities, such as traveling to and from school, children construct place meanings and attachment through seeing and interacting with bicycle paths and streets, apartments and businesses, and other children and adults (Chawla 1994; Larsen and Harrington 2018). Note that although they may already have developed meanings related to a place (e.g., green spaces, violence), youth may enter a program with little place attachment. This is more likely if they live in neighborhoods with weak social cohesion and signs of decline such as trash, vacant lots, and unruly behaviors, or otherwise have had negative experiences with their place (Brown et al. 2003). Participants who have moved to a new city or experienced a disaster may also lack feelings of place attachment (Ardoin 2006). Thus, educators will want to consider participants' existing sense of place in planning program activities. Similar to the situation with attitudes, educators can "navigate" existing place meanings, attachments, and identities (Heberlein 2012; see also chapter 7), while also taking steps to enhance ecological place meanings and related place attachments.

Because gender, ethnicity, and class influence how people use places, which in turn determines their sense of place, educators will also want to use their knowledge of the sociopolitical context in which their participants live to plan activities (cf. Uzzell et al. 2002; Manzo and Perkins 2006). In some cases, a minority ethnic group may experience higher levels of place attachment than the majority group, perhaps as a result of feeling at home in a neighborhood with fellow minorities (Brown et al. 2003). Further, in urban neighborhoods, residents may express their cultural identity through creating "sacred places" and "vernacular" spaces that have local significance, such as community gardens with ethnic plants and folk art (Manzo and Perkins 2006). Once having identified such places, environmental educators can help participants describe associated place meanings, through visiting, mapping, drawing, photographing, videoing, and conducting stewardship activities in these places. When sacred places are threatened by development, participants can communicate their place meanings and attachments to local decisions makers in an attempt to preserve them (Hester 1993). Finally, when program participants disagree about how a place should be treated, educators can ask them to discuss their underlying place meanings and place attachment (Manzo and Perkins 2006).

In addition to taking into account existing sense of place, educators can engage participants in activities to build place attachment and ecological place meanings. These activities should be enjoyable and conducted with other children and adults. For children, spending unstructured time exploring nature is the primary means for developing ecological place meanings and attachment (Chawla 1994; Briggs et al. 2014). Positive experiences in nature also can be created through nature-based recreation (Larson et al. 2018), community gardening and similar civic ecology practices (Krasny and Tidball 2015), and community celebrations (Brown et al. 2003). Another approach to cultivating place attachment is having long-term residents who are deeply connected to a particular neighborhood or place share their oral history with newcomers or youth (Brown et al. 2003; Stefaniak et al. 2017). This approach can piggyback on popular storytelling activities (e.g., Story Maps, StoryCorps). Such intergenerational programs also create social cohesion, which is needed in order for sense of place to result in community action (Uzzell et al. 2002).

In trying to foster sense of place among participants, educators will want to consider that while place meanings may change through environmental education programs, place attachment develops through a series of interactions over a long period of time. In programs in the Bronx, youth who had not previously spent time along the Bronx River and had place meanings focused on the built environment participated in summer stewardship and monitoring activities along the river. After participating in the programs, they developed an ecological place meaning, now viewing the Bronx as a place where they could see wildlife and enjoy outdoor recreation (Kudryavtsev et al. 2012). In these programs, nurturing ecological place meaning involved opening participants' eyes to the ecological dimensions of their neighborhoods that they had not noticed before. However, the youths' place attachments did not change as a result of the programs, suggesting that multiple experiences with places and people over a longer period of time may be needed to develop place attachment. Such long-term experiences should cultivate feelings of satisfaction, security, and pride in one's place, and foster meaningful social interactions (Stedman 2002; Lewicka 2011). Even if some students are already strongly attached to their community, an environmental education program might influence the reason for this attachment through nurturing ecological place meanings (Kudryavtsev et al. 2012).

Nature-based recreation can also provide a pathway to sense of place and related conservation behaviors (Larson et al. 2018). People who spend time in urban parks connect to the park and to the people who use the park (Peters et al. 2010). Sites where participants engaged in bird-watching or hunting were associated with emotional components of place attachment, including positive memories and a sense of confidence and comfort. Additionally, different types of recreation may be linked with distinct ecological or social place meanings and identities. Bird watchers hold ecological place meanings, whereas hunters' place meanings are defined by social interactions in rural communities and a hunting culture (Larson et al. 2018).

Citizen science projects can be another approach to developing sense of place, particularly those that entail repeatedly going back to the same place to collect data on biodiversity. During these repeated activities, participants learn about the data-collection site and develop place meanings, and form ties with fellow data collectors and with the place. In a citizen science project focused on coastal seabirds, participants attributed becoming attached to the place where they collected data to the time and effort they put into data collection, the beauty of the site, encountering wildlife, and becoming familiar with the site and related scientific knowledge (Haywood et al. 2016).

Other activities that enhance place meanings and place attachment include action research, community service learning, discussions with peers and environmentalists, place-based interpretive excursions, nature studies, and stewardship of a local park or other resource (Kudryavtsev et al. 2011; Russ et al. 2015; Adams et al. 2017). Students might also engage in concept mapping to highlight important places and networks, for example those related to food and energy sources or recreation. When discussing their maps, students can recognize how their activities connect to the larger city, and can reflect on issues of power and equity in relation to air and water contamination and access to green space (Adams et al. 2017). Regardless of the particular activity, consider the potential for participants, their families, and the broader community to work together, form positive memories, express themselves, and develop respectful relationships (Chawla 1994; Payton et al. 2005; Smith and Sobel 2010; Johnson et al. 2012; Adams et al. 2017).

Place-based education is a popular approach in environmental education. Based on the belief that students and community members should first understand concerns within their local community prior to tackling national and international problems, it engages participants in action research and other activities in the local community (Sobel 2004). More broadly, place-based education is concerned with raising awareness of place, our relationship to place, and our positive contributions to place, to environmental quality, and to community well-being (Adams et al. 2017). Often it is used as a means to teach school subjects across the curriculum and improve academic achievement (Sobel 2004; Smith and Sobel 2010; Johnson et al. 2012). By providing repeated experiences with community members, place-based education should enable students to develop place meanings and ties to their community and environment (Larsen and Harrington 2018).

Educators wishing to incorporate a sociopolitical perspective into place-based learning can draw on critical pedagogy (Freire 1973). Critical pedagogy of place integrates place-based learning with learning about the sociopolitical forces, such as inequality, that impact places and their inhabitants. It entails decolonization, or "learning to recognize disruption and injury and to address their causes," followed by reinhabitation, or "learning to live well socially and ecologically in places that have been disrupted and injured" (Gruenewald 2003, 9). Such perspectives integrate awareness of program participants' social, cultural, and political context with efforts to build sense of place.

Assessing Sense of Place

Multiple methods are used to assess sense of place, including activities embedded in environmental education programs. Participants can take photos of significant, favorite, or even disliked places as part of an activity to explore their neighborhood. During a group discussion or one-on-one interviews afterward an evaluator can ask participants to explain why they took photos of certain places and what those places mean to them (Briggs et al. 2014). Such photo-elicitation methods can be expanded using computer technologies, for example by allowing children to pin their photographs to a digitized map or record stories about their favorite places. Alternatively, children might use a printed map of significant places in their neighborhood or simply be asked to draw a map of significant places (Lewicka 2011). Regardless of the method used, paying close attention to the meanings participants ascribe to places, plants, wildlife, and other features is critical, as participants' associations with places may differ from those of evaluators (Briggs et al. 2014).

Likert scale surveys are also commonly used to assess participants' sense of place, with questions that reflect place meaning and place attachment (Kudryavtsev et al. 2011; Kudryavtsev et al. 2012). Place-meaning questions should include items about ecological features of the place, whereas place-attachment questions can capture both place identity and place dependence (see appendix).

10

EFFICACY

Highlights

- Efficacy refers to people's beliefs about whether their actions will achieve their individual or group goals.
- Self-efficacy and collective efficacy influence individual behaviors and collective action, and political efficacy influences whether we engage in the policy-making process.
- Efficacy influences our environmental behaviors through its impact on the goals individuals and groups set for themselves, actions they are willing to take, and their perseverance in overcoming obstacles to achieve desired outcomes.
- Environmental educators foster efficacy through providing participants with mastery experiences, role models, and supportive social interactions, and paying heed to participants' emotions.

When we believe we will reach our goals—as individuals, groups, or part of a political system—we are more likely to take action. Mahatma Gandhi captured this idea in saying, "If I have the belief that I *can* do it, I shall surely acquire the capacity to do it" (quoted in Deats 2005, 108).

What Is Efficacy?

> **Unless people believe they can produce the desired effects by their actions, they have little incentive to persevere in the face of difficulties.**
>
> (Bandura 2004, 79)

Individuals have self-efficacy, whereas groups have collective efficacy. We can also have political and civic efficacy, which are useful in predicting whether we vote or engage in other civic behaviors. People with participative efficacy believe that their contributions to a larger cause, like reducing plastics pollution, matter. In short, psychologists, sociologists, and even political scientists investigate how different types of efficacy influence a suite of environmental behaviors and collective actions (table 10.1).

Self-efficacy is a belief that one can succeed in a specific situation or accomplish a task. Our self-efficacy plays a major role in how we approach goals, tasks, and challenges. Similarly, our belief that we can accomplish something strongly affects whether we attempt it and eventually succeed (Bandura 1993). Self-efficacy is often invoked in studies of academic achievement (Bandura 1993) and personal health (Strecher et al. 1986), where a person's behaviors directly impact desired outcomes. However, environmental problems require more than the behavior of one individual to solve; thus other forms of efficacy come into play.

Collective efficacy shifts the focus from individual to group goals and actions. It is the belief that the problems of a group can be solved through collective activity (Van Zomeren et al. 2013; Barrett and Brunton-Smith 2014), or a shared "belief in the capacity of the group to pull together and realize shared aspirations or address shared problems" (Watts and Flanagan 2007, 786). A neighborhood's or group's collective efficacy predicts whether they will engage in collective action to benefit the greater good. For example, working in Chicago in the 1990s, researchers found that residents who felt a sense of collective efficacy were likely to take actions to address neighborhood problems—like transforming a trash-strewn lot into a community garden or reprimanding teenagers engaged in unruly behavior (Sampson et al. 1997; Hurley 2004).

Although collective efficacy can help explain why a group engages in collective action, it fails to explain why any one individual chooses to participate in that action. This is known as the free-rider problem (Ostrom 1990). For example, I might believe that my neighborhood is fully capable of converting a vacant space to a park and, thus, that my neighborhood has collective efficacy to reach this goal. But I ask myself, why should I participate when everyone else is fully capable of accomplishing the work? (In other words, why contribute when I can "free ride"?) Here is where a less discussed but important form of efficacy—*participatory efficacy*—comes in. Participatory efficacy is a belief that my own individual actions in converting that park will make a difference, or more generically, a belief that one's own contribution to a group's action matters. Young people may believe that they have more influence on social problems, and thus higher participative efficacy, relative to adults (Fernández-Ballesteros et al. 2002; Van Zomeren et al. 2013).

Political efficacy refers to "the belief that political change is possible and that we have the capacity to contribute to it through deliberate judgments and actions" (Beaumont 2010, 525). It can be considered as a type of collective efficacy or beliefs about whether a group or the broader public has the ability to effect change in the political process (Lee 2006; Beaumont 2010). People have both internal and external political efficacy. Internal political efficacy is the belief that one has the ability to participate in the political process, whereas external political efficacy is the belief that government is responsive to citizens' demands (Barrett and Brunton-Smith 2014). Internal political efficacy is often considered as a type of self-efficacy referring to an individual's perception about his or her own potential to impact the political process. Closely related to political efficacy is *civic efficacy*, which reflects our beliefs about the effectiveness of participation in civic life (Serriere 2014). For students, who may have little opportunity to influence policy outside of school, *school efficacy* reflects similar beliefs about one's ability to impact school policies (Torney-Purta et al. 2001).

TABLE 10.1 Types of efficacy

TYPE OF EFFICACY	DEFINITION
Self-efficacy	Belief in one's ability to succeed in specific situations or accomplish a task (Bandura 1977)
Collective efficacy	Group's shared belief in its collective abilities to organize and execute the courses of action required to reach goals (Bandura 1997)
Participative efficacy	Belief that an individual's contribution is important to the success of collective action (Van Zomeren et al. 2013)
Political efficacy	Belief in our abilities to understand the political realm and act effectively in it (Beaumont 2010)
Internal political efficacy	"Belief that one understands civic and political affairs and has the competence to participate in civic and political events" (Barrett and Brunton-Smith 2014, 15)
External political efficacy	"Belief that public and political officials and institutions are responsive to citizens' needs, actions, requests, and demands" (Barrett and Brunton-Smith 2014, 15)
Civic efficacy	Belief that one's actions can make a difference in the civic life of one's community (Serriere 2014)
School efficacy	Belief that actions taken by groups of students can improve their school (Torney-Purta et al. 2001)

Why Is Efficacy Important?

> **Unless people believe they can produce desired outcomes and forestall undesired ones through their actions they have little incentive to act or to persevere in the face of difficulties.**
>
> (Fernández-Ballesteros et al. 2002, 108)

Simply stated, efficacy is important because if we don't believe we will succeed, we are unlikely to try.

- Self-efficacy is related to environmental behaviors in adolescents (Meinhold and Malkus 2005) and to collective actions in adults (Lubell 2002).
- Collective efficacy predicts an individual's plans to engage, and actual engagement, in environmental behaviors (Homburg and Stolberg 2006; Barrett and Brunton-Smith 2014; Chen 2015; Barth et al. 2016; Jugert et al. 2016). Collective efficacy is especially important in addressing environmental issues, where in contrast to other endeavors like studying harder to improve test scores, individual behaviors do not result in immediate observable outcomes.
- Collective efficacy predicts collective environmental action (Homburg and Stolberg 2006; Chen 2015; Barth et al. 2016).
- Collective efficacy can help alleviate feelings of helplessness or hopelessness given the enormity of environmental problems relative to actions we can take as individuals (Meinhold and Malkus 2005; Chen 2015; Barth et al. 2016).
- Political efficacy is one of the strongest predictors of people's political participation, including whether or not an individual is likely to vote, contact representatives, or become involved in political activism. Internal political efficacy appears to have a more consistent effect on different types of political participation relative to external political efficacy (Beaumont 2010; Levy 2013; Barrett and Brunton-Smith 2014).

Author Reflections

Efficacy—in its multiple forms—has had a profound impact on my life. The summer of my seventeenth birthday, I worked at a camp near Mount Rainier in Washington State. I remember on his days off, one of the boys working at the camp took a glacier-climbing class and summited Mount

Rainier. The camp boss would not allow me to take the class because I was a girl. I truly believed I did not have the capacity to climb Mount Rainier. I also believed for years I could never finish a marathon, but when I tried for the first time in my early sixties, I completed the race, and two years later, I qualified for the Boston Marathon. Perhaps I am lucky that my self-efficacy has grown over the years.

After President Trump won the election, both internal and external political efficacy have crept into my life. I participated in the Women's March in Washington, DC, and March for Science in New York City. These big protests, though perhaps not impacting policy immediately, spawned a new group of activists across the US, who believe that they can change the system and that the system can be changed. In short, they have internal and external political efficacy. Many of these women—as well as men spurred to action by our president—were just elected to Congress. I have joined this movement in a smaller way—by knocking on doors to try to persuade people to elect congressional candidates who support environmental policies. Early on, one of the candidates I supported lost a primary, but this week, one of the House candidates I volunteered for won. Even though I believe I played a minuscule part in his victory (had little participative efficacy), the fact that he won has reinforced my sense of internal political efficacy. Having victories, whether personal, collective, or political, seems critical to building and reinforcing efficacy of all kinds. In daily life and in environmental action, the feedbacks between efficacy and action—how they build on each other in an iterative fashion—are on display.

How Does Efficacy Contribute to Environmental Behaviors and Collective Action?

> **It seems that being a member of a group changes our beliefs about what we can achieve. Even though we ourselves cannot solve pressing problems such as climate change we may feel that as a group we have the power to make a difference.**
>
> (Barth et al. 2016, 66)

Self-efficacy impacts environmental behaviors through several pathways, including by influencing a person's goals, aspirations, and motivations; what actions a person decides to take and how much time, energy, and persistence to put into that action; and a person's emotional ability to deal with environmental issues (Bandura 1993; Lubell 2002; Young 2017). Self-efficacy is closely tied to locus of control—that is, individuals' perceptions of their ability to bring about change through their behavior. Locus of control in turn has been linked to environmental behaviors through empowering individuals, or providing a sense that one can make changes and help resolve environmental issues (Hines et al. 1986/87; Hungerford and Volk 1990; Ernst et al. 2017).

Collective efficacy helps people who fear that individual behaviors do not make a difference envision how they can impact the environment by joining with others (Homburg and Stolberg 2006; Barth et al. 2016). Individuals realize that when people provide mutual support, form alliances, and pool their knowledge, skills, and resources, they have a greater ability to achieve a desired outcome (Fernández-Ballesteros et al. 2002). In addition, when a group shares a sense of collective efficacy, members may increase their confidence in their own ability to make change; in this way, collective efficacy contributes to individuals' self-efficacy (figure 10.1; Jugert et al. 2016).

As a reflection of our beliefs about the effectiveness of our actions, political and civic efficacy affect our willingness to engage in political and civic behaviors (Barrett and Brunton-Smith 2014). Participative efficacy similarly reflects beliefs

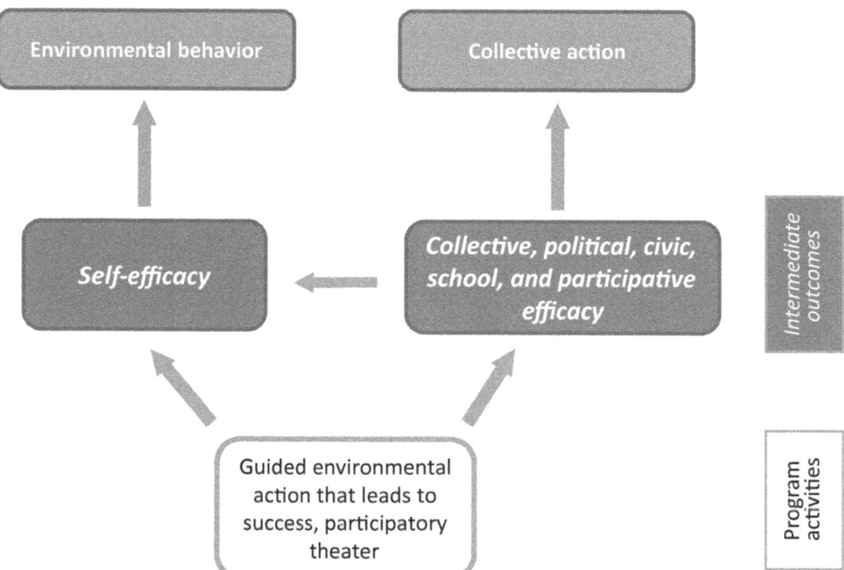

FIGURE 10.1. Self-efficacy predicts individual environmental behaviors, whereas various forms of group efficacy are tied to collective action. Collective efficacy can also lead to self-efficacy.

about the contributions of our individual actions and thus impacts decisions about joining in collective action (Van Zomeren et al. 2013).

How Can Environmental Education Build Efficacy?

Environmental educators can use four general strategies to foster self-efficacy and political efficacy in youth: mastery experiences, vicarious experiences, social interactions, and paying heed to participants' emotional state (Bandura 1977; Beaumont 2010). A group's collective efficacy depends on having successfully worked together in the past to achieve a common goal and to effect social change. Thus, environmental educators can enhance collective efficacy by engaging program participants in collective environmental actions (see chapter 5) in which they depend on each other to achieve a shared goal (Fernández-Ballesteros et al. 2002; Velasquez and LaRose 2015).

Mastery Experiences

Mastery experiences allow youth to master a skill or behavior. Such experiences can foster self-efficacy when they include authentic hands-on opportunities to practice a specific behavior, to experience how the world responds to one's efforts, and to achieve success. For small children, playing in natural areas can provide mastery experiences. For example, when children build a dam with rocks in a small stream, they can see how the water changes course—the result of their actions (Chawla 2006). Older children and adolescents develop self-efficacy through more challenging mastery experiences guided by experienced and supportive adults. Breaking down behaviors or actions into simple, achievable steps, and steering youth away from overambitious efforts where their limited abilities or structural barriers are likely to result in failure, are also important to building self-efficacy and other forms of efficacy. Once students master relatively easy yet still challenging behaviors and build a sense of efficacy, they may feel empowered to take on more challenging behaviors (Lauren et al. 2016; Reese and Junge 2017). Mastery experiences are the most important strategy for fostering efficacy because they allow students to actually practice the behavior and enable them to see their accomplishments (Bandura 1977).

Vicarious Experiences

Vicarious experiences have to do with seeing others—usually role models—take action and achieve success. Students can observe, interact with, and emulate role models who are knowledgeable about policy processes and are politically involved (Beaumont 2010). A teacher who rides his bike to school might inspire students to do the same and might even help those students to advocate with city council for bike lanes. Students also benefit from outside speakers, such as local politicians or community members, who are inspiring and have compelling stories of overcoming challenges (Beaumont 2010).

Social Interactions

Social interactions include expressing faith in students' abilities, realistic encouragement (rather than empty pep talks), persuasion, supportive relationships, access to professional and community networks, and actual inclusion in a political community (Bandura 1977; Beaumont 2010). Such interactions entail guiding students through an activity—letting them know when they are succeeding and when they need to reflect and consider an alternative strategy. This includes providing feedback, opportunities for reflection and discussion, as well as helping students connect to groups that value and support their interest in influencing policy. Especially important is developing a sense of community with peers and community members engaged in the policy process, for example through internships and service learning (Beaumont 2010).

Emotions

In fostering efficacy, educators need to be attentive to students' emotional state. A student who is anxious about being in a program, or is feeling vulnerable or out of place, is less likely to develop feelings of self-efficacy (Bandura 1977). An environmental educator should be aware of students who may feel stress or anxiety and help them become comfortable with the group and its activities. Further, environmental educators need to provide a basis for hope tempered with reality, demonstrate the power of working as a group rather than alone, and help students to persist through the inevitable challenges that arise (Jugert et al. 2016). Finally, educators can help students appreciate the value of the "sense of dignity, community, and solidarity that can come from an active political life" (Beaumont 2010, 554).

Implementing Efficacy-Building Strategies

Given environmental education's history of engaging youth in hands-on experiences, it is easy to imagine how environmental education programs might incorporate mastery experiences, role models, and social interactions, and pay heed to program participants' emotional state. Environmental action (see chapter 5) and positive youth development approaches to environmental education (see chapter 14) in particular incorporate multiple strategies for building self-, collective, and political efficacy.

In the "Salad Girls" program, elementary school students decided they wanted to change the school lunch program to reflect their faith-based dietary restrictions. Their teacher helped create a "mastery experience" for the girls. She connected her lessons to her students' interests and experiences, encouraged student inquiry, valued her students' ethnic and religious diversity, and allowed students to practice activism skills (Serriere 2014). The cafeteria worker also supported the students' desire to change the lunch offerings, thus alleviating a structural barrier that could have hindered the students' success in building efficacy and achieving their goals.

Engaging youth in school-based civic literacy programs can enhance political and civic efficacy. In particular, discussions of political issues and opportunities to build rapport with politically engaged peers can build a classroom culture of political interest and strengthen students' political efficacy. Further, teachers can help students build trust in government and civil society institutions and encourage students to persist in achieving their goals for social or political change as they face challenges over an extended period of time. This may involve suggesting alternative pathways for achieving change when an initial strategy fails (Levy 2013). Importantly, environmental education can provide real-life experiences in the community for students in civics classes.

Narrative and participatory theater also have been used to build self- and collective efficacy, including with vulnerable groups such as women and those living in developing countries. These methods draw on and validate participants' traditional storytelling practices, while providing opportunities to act out new roles and form social connections to support collective action. Educators also can incorporate opportunities for discussions, teamwork, and self-evaluations into narrative and participatory theater (Young 2017).

Assessing Efficacy

Researchers use "I" statements to measure self-efficacy and "we" statements to measure collective efficacy. Likert scale survey statements are worded either more generally about the environment or specifically about a particular type of environmental behavior or action (see appendix).

For general self-efficacy, survey statements might be worded as "I am optimistic that I can protect the environment" or "I am capable of protecting the environment" (Reese and Junge 2017). A specific measure of self-efficacy in a study of plastics behaviors asked participants their level of agreement with the statement

"I think that I am capable of protecting the environment by means of my personal plastic reduction" (Reese and Junge 2017). Similar questions were used before and after a training program for farmers impacted by drought in Malawi: "I feel confident that I can deal with unexpected events"; "When confronted with a problem, I can usually find several solutions"; and "I can solve most problems if I invest the necessary effort" (Young 2017).

To measure general collective efficacy, evaluators can ask participants their level of agreement with the statement "People in the community can come together to solve problems" (Young 2017). In studies investigating climate change actions, more specific collective efficacy was measured by level of agreement with the following statements: "Climate change can be averted by mobilizing collective effort"; "If we act collectively, we will be able to minimize the consequences of climate change" (Morton et al. 2011); and "As inhabitants of this region we can do much to noticeably reduce CO_2 emissions together" (Barth et al. 2016). Researchers also have asked questions about participatory efficacy to determine individuals' perceptions of the importance of their individual contributions to collective action. For example, "I believe, as an individual, that I can contribute to students' ability to change energy policy in my school" (Van Zomeren et al. 2013). Finally, you may want to measure internal and external political efficacy (see appendix), or, in situations where students have little opportunity to influence policy, you can measure school efficacy by posing survey statements such as "Lots of positive changes happen in this school when students work together" (Torney-Purta et al. 2001).

You can also use open-ended interview questions to gain in-depth understanding of participants' efficacy before and after an environmental education program. For self-efficacy, participants might be asked to "describe your ability to impact environmental problems." For collective efficacy, you might ask, "What challenges do you see in working as a community in responding to changes in the weather in the past few years?" (Young 2017).

11

IDENTITY

Highlights

- Identity refers to the labels we give to ourselves, the groups we belong to, and how we distinguish ourselves from others.
- We all have multiple identities that may change or become more salient over time and that influence how we interpret information.
- Ecological, environmental, place, civic, and collective identities can influence our individual behaviors and collective actions.
- The activities we engage in influence our identities and their salience.
- Environmental education, including approaches that engage participants in environmental action, debates, and developing a sense of responsibility for nature, can foster environmental identity.

We all have multiple identities—as an environmentalist, a Muslim, a man, a conservative. Some of these identities are largely fixed or imposed from the outside (e.g., gender, race), while others we choose, like environmentalist. Identities can also change over time as a result of new information and experiences. For example, volunteering for a litter cleanup, or participating in a debate about climate change, can foster an environmental identity. Thus, performing an environmental behavior or collective action contributes to our environmental identity, which in turn can make us more likely to engage in additional behaviors and actions.

Identities influence how we interpret information, as well as our behaviors and collective actions (Kitchell, Kempton, et al. 2000; Clayton 2012; Fielding and Hornsey 2016).

What Is Environmental Identity?

> **Identity is fundamentally a way of defining, describing, and locating oneself. . . . People have multiple identities that can vary in salience and significance over a lifetime and across different contexts.**
>
> (Clayton 2012, 165)

Identity refers to how we label ourselves, how other people label us, and how we distinguish ourselves from other individuals and groups. Personal identities are unique to ourselves as individuals, whereas we share social identities with others (table 11.1).

Individuals who spend time in nature often form an *ecological identity*, which refers to how we view ourselves "in relationship to the earth as manifested in personality, values, actions, and sense of self" (Thomashow 1995, 3). Whereas the definition of *environmental identity* is similar to that for ecological identity, environmental identity focuses more on environmental behaviors and actions and less on spending time in nature. Environmental identity refers to "a sense of connection to some part of the nonhuman natural environment that affects the way we perceive and act toward the world; a belief that the environment is important to us and an important part of who we are" (Clayton 2003, 45–46). Environmental identity can be a personal identity, for example our identity as a thoughtful consumer or vegetarian, and can predict individual environmental behaviors (Nigbur et al. 2010). It can also be a social identity, as when we are part of an environmental movement (Kempton and Holland 2003; Holland et al. 2008; Dono et al. 2010). Because social identities revolve around a shared understanding of group norms and goals (Tajfel and Turner 1986), they can be used to mobilize action and to exclude others. For example, hunters, African Americans, and white environmentalists may share a concern about and work collaboratively to protect the environment, but the way in which white environmentalists define environmental identity may make other groups feel excluded (Holland and Lave 2009).

Whereas environmental identity motivates action to protect the environment more broadly, place identity motivates people to protect a particular place (Stedman 2002; Clayton 2012; see also chapter 9). *Place identity* includes our memories, ideas, feelings, and other cognitions about a physical setting and the social interactions that occur in that setting (Proshansky et al. 1983; Uzzell et al. 2002). Similar to environmental identity, place identity can be either an individual (Proshansky et al. 1983) or shared, social identity (Uzzell et al. 2002). We often think about our local place identity; for example, when people ask where are you from, you might respond by naming the city where you live. However, place identities can occur at different scales, including neighborhood, city, region, and country (Lewicka 2008). When a place is central to our identity and we value its natural features, we are more likely to support preserving that place (Stedman 2002; Carrus et al. 2005; see also chapter 9).

Collective identity focuses on how group members contribute to social movements (Polletta and Jasper 2001; Saunders 2008) and how they act in a commons dilemma (De Cremer and Vugt 1998). If you were faced with a commons dilemma, you would have to choose between your own interest, which would lead to benefits for you in the short term, and conserving a common resource for the long-term benefit of everyone in your community. So, for example, if you are a farmer, utilizing the greatest amount of irrigation water helps you grow crops, but if all the region's farmers withdraw the maximum groundwater for their own farm, eventually the groundwater dries up, and all suffer. If you share a collective identity with other farmers, and if you recognize that other farmers share the common groundwater resource, you are more likely to trust, care about, and cooperate with fellow farmers (Dovidio et al. 2007). Researchers have also shown that in situations where one's collective identity is made more salient relative to one's personal identity, people are more likely to support a public good such as a city park (De Cremer and Vugt 1998).

Similar to collective identity, *civic identity* is linked to trusting and helping fellow community members, and working toward a common goal or public good. Civic identity consists of a sense of belonging to a community and having rights and responsibilities related to that community, including contributing to its well-being (Atkins and Hart 2003).

Regardless of whether we incorporate nature, the environment, a place, our community, or civic and other values into our identity, we all balance multiple identities when we decide to change our behavior or take environmental action. For example, we might balance our identities as an athlete, an environmentalist, a person of faith, and an immigrant in deciding which volunteer activities to join. Once we have chosen certain activities, they tend to reinforce the related identity. Choosing to play sports rather than joining an environmental club will reinforce a student's identity as an athlete (unless she has negative experiences playing sports, in which case she may stop identifying as an athlete and seek another activity and related identity).

TABLE 11.1 Types of identity

TYPE OF IDENTITY	DEFINITION
Individual identity	How we define ourselves (e.g., mother, Christian, environmentalist, nurse)
Social identity	Aspects of self-image that derive from the social categories (e.g., environmentalist, millennial) to which we perceive ourselves as belonging (Tajfel and Turner 1986)
Ecological identity	How we view ourselves "in relationship to the earth as manifested in personality, values, actions, and sense of self" (Thomashow 1995, 3)
Environmental identity	"Sense of connection to some part of the nonhuman natural environment that affects the way we perceive and act toward the world; a belief that the environment is important to us and an important part of who we are" (Clayton 2003, 45–46). Can be individual or social (e.g., part of an environmental movement).
Place identity	Identity related to memories, ideas, attitudes, and feelings about a physical setting and the social interactions that occur in that setting (Proshansky et al. 1983; Jorgensen and Stedman 2001)
Collective identity	Group identity focused on members' associations with and contributions to social movements and their organizations (Polletta and Jasper 2001; Saunders 2008)
Civic identity	Feeling of connection to a community and having rights and responsibilities related to that community (Atkins and Hart 2003)

Why Is Identity Important?

Identities play a critical role in determining what information we accept and what actions we take. While environmental and related identities foster environmental behaviors and collective actions, conservative political identities play a role in people's questioning of climate change science and lack of support for climate and other environmental policies.

- Identities impact who and what information we dismiss or pay attention to, in part by determining our emotional reactions to messengers and messages (Clayton 2012).
- Identities influence our behaviors and actions (Clayton 2012).
- Identities appear to be more effective predictors of a broad spectrum of environmental behaviors than attitudes that focus more narrowly on specific behaviors (Gatersleben et al. 2014).
- If we fail to consider our audiences' identities, we may use ineffective messages and even incite our audiences to non-environmental behaviors to protect an identity they feel is threatened.

Identity (Adapted from Clayton 2012)

Identity is a label that we use to define, describe, and locate ourselves within society.

Identities describe our personal attributes, whom we are like and unlike, and what groups we are tied to.

Identities develop over time and within particular social contexts.

We have multiple identities that differ in their strength or salience.

Some identities, like "environmentalist," are chosen and can change over time. Other identities, like ethnicity, are fixed.

We adopt particular identities to fill needs, like the desire for belonging or self-esteem.

We filter information according to our identities.

We choose actions according to our identities.

How salient an identity is at a particular time predicts which behaviors or actions we engage with.

Sometimes we use our identities defensively—to resist influence or threats.

Identities are more powerful predictors of behavior or collective action than knowledge or attitudes.

If we create opportunities for program participants to experience belonging and self-esteem in an environmental education program, they are more likely to develop an environmental identity.

How Does Identity Contribute to Environmental Behaviors and Collective Action?

Identity contributes to individual behaviors and collective action by influencing the way we process information, how we view others, and the social norms we subscribe to (figure 11.1). It can also impact behaviors by engendering positive and negative emotions.

Social identities affect the way we process information. For example, we are more likely to pay attention to and trust information from someone who shares our political identity as a liberal or conservative. Often educators act on the belief that facts change people's behavior. But what happens when our identities conflict with the facts we have been taught? In this case, we often interpret new information, regardless of its validity, in ways that reflect the beliefs of the groups to which we belong. The notion that our social-political identities can trump what we know about science has been used to explain the differences in how liberals and conservatives with similar science knowledge support different climate policies (Kahan 2015).

Social identities also influence behaviors and actions through the norms or expectations of the groups with which we identify. For example, you may identify as an environmental educator and feel pressured to bring a reusable water bottle to conferences, to buy carbon offsets to compensate for your flight emissions, and to volunteer for a fund-raiser because you perceive these behaviors as norms for environmental educators. Social identity theory helps explain these behaviors. It suggests that once we categorize ourselves as belonging to a particular group, we start to accentuate our similarities with others in our group while emphasizing our differences with people in other groups. We then assimilate our attitudes, beliefs, and behaviors to the norms of our group while polarizing away from norms of other groups (Tajfel and Turner 1986; Fielding and Hornsey 2016; see also chapter 12). Whereas adopting group norms may be helpful in spurring environmental behaviors, this polarization can thwart progress in resolving environmental conflicts or advancing environmental policy.

Emotion influences how identities interact with behaviors (Devine-Wright and Clayton 2010). When others confirm our identities, we feel better about ourselves. But when we sense our identities being threatened, negative emotions

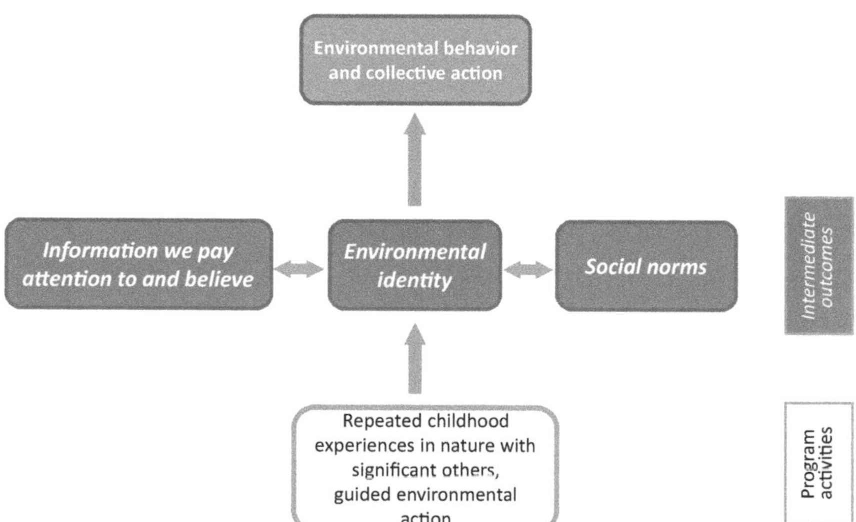

FIGURE 11.1. Environmental identity impacts environmental behaviors and collective action by determining what information we pay attention to and believe and what group norms we adopt.

ensue, which can result in our reconsidering how we define ourselves, blaming others for our behaviors, or sticking more firmly to our original identities. Terror management theory explains how people who do not identify as environmentalists might have emotions that lead to non-environmental behaviors (Dickinson 2009). Because identities are tied to our self-image and social standing (Clayton 2012), when people who do not identify as environmentalists are confronted with a doom and gloom environmental message, they may feel that their self-image is threatened, and they may even be reminded of their mortality. In response, they may decide to undertake behaviors that make them feel more important and less vulnerable—such as driving fast in a large SUV. In short, people sometimes manage feelings of mortality or "terror" that environmental messages incite by engaging in non-environmental behaviors (Dickinson 2009).

Environmental contamination and disasters can also threaten our social or place identity, leading to feelings of hopelessness or, alternatively, to collective action. Hopelessness occurs in part when we hear negative descriptions of our community, develop a sense of insecurity, and feel as if we have limited control over our life (Edelstein 2002). Sometimes those impacted by a contamination event like the BP oil spill in the Gulf of Mexico react by taking collective action, such as cleaning oil-covered seabirds or monitoring long-term water quality. In so doing, they may form new collective or social identities tied to place and to being an environmental citizen (Clayton 2012).

People joining an environmental group may not initially describe themselves as environmentalists but adopt an environmental identity through active participation in environmental activities (Kitchell, Kempton, et al. 2000). In one study, zoo volunteers adopted an identity as a zoo volunteer, which became important to their self-esteem, provided social support, was recognized by others, and gave them a sense of purpose and of self-efficacy. This led to them becoming active in other environmental activities outside the zoo (Fraser et al. 2009). In short, there is a strong feedback loop between behaviors and identity, with identity impacting behaviors and behaviors impacting identity.

Author Reflections

Although I have written largely about environmental identity in this chapter, I think about social, cultural, and political identities all the time. Americans' social, cultural, and political identities seem to have the most influence on what the US is doing—or not doing—about the existential threat of climate change. More and more conservatives are accepting the fact that the climate is changing—largely owing to the difficulty of ignoring the evidence they see before their eyes, whether it is devastating floods or catastrophic fires. Yet our beliefs about what the government should do about climate change are still determined by the "tribes" to which we belong. Because people hold strong social identities that are not likely to change, some of the most creative efforts I have observed find superordinate identities that both social conservatives and liberals share. For example, conservatives and liberals living along the Virginia coast may share an attachment to their common place, or have a shared place identity. By invoking that shared place (or superordinate) identity, environmental educators can bring different groups of people together in stewardship actions.

How Can Environmental Education Nurture Environmental Identity?

> **To achieve the aim of sustained action for the environment, environmental education needs to provide a pathway through a series of social settings where young people will want to belong, where their identity as an environmental actor will progressively "thicken."**
>
> (Williams and Chawla 2016, 993)

Repeated childhood experiences in nature with a caring adult can lead to a lifelong environmental identity (Chawla and Cushing 2007). How might nature become part of one's identity? Nature elicits strong emotions and memories while allowing us the freedom to be ourselves without judgment from others. When we succeed in a challenge presented by nature, such as reaching the summit of a mountain or growing a garden, our self-esteem and self-efficacy are enhanced, making us want to engage in additional nature-based activities. Spending time in nature also can be a means of demonstrating our social and political commitments and can spur social connections, thus satisfying a need to belong to a group (Clayton 2012; Williams and Chawla 2016). In short, nature is a setting for strong emotions, feeling good about ourselves, demonstrating our commitments, and forming connections with others, all of which contribute to an environmental identity.

Additionally, service learning (Stapleton 2015), experiences that include a focus on environmental problems in an environmental sciences classroom (Blatt 2013), and programs where students develop a sense of responsibility for nature through modeling by teachers, foster an environmental identity (Williams and Chawla 2016). For adults, engagement with different types of environmental organizations, such as local stewardship or environmental justice nonprofits, influences specific environmental identities and actions (Kitchell, Kempton, et al. 2000; Saunders 2008). Further, storytelling describing environmental problems and actions taken to address them can help build and reinforce an environmental identity (Kitchell, Hannan, et al. 2000). Finally, because individuals balance multiple identities in deciding to take action, and identities change over time based on new experiences, repeated experiences are important to maintaining and making salient environmental and collective identities (Polletta and Jasper 2001; Williams and Chawla 2016).

Four stages in developing an environmental identity were found in a program for secondary students (Stapleton 2015). First, temporarily living alongside people who had suffered severe flooding helped students recognize the *importance* of environmental issues. Upon returning home, the students designed local action projects, through which they began to *see themselves as individuals who take environmental action* and *became knowledgeable about how to engage in environmental action*. Finally, *recognition of their environmental action and knowledge by parents, teachers, and peers* was critical to students forming an environmental identity (Stapleton 2015). In young children, nature experiences that build children's trust in their natural environment, enable autonomy and initiative, and build a sense of accomplishment foster an environmental identity (Green et al. 2015).

Three factors help people develop a civic identity: volunteering, service, or other forms of community participation; learning about one's community; and a commitment to fundamental democratic principles such as justice and fairness (Atkins and Hart 2003). However, the ability of children and adults to access volunteer, service learning, and other experiences that help create an environmental or civic identity may be limited among low-income individuals, communities of color, and immigrants (Flanagan and Levine 2010; Chung and Probert 2011) as well as in countries where civil society activity is curtailed. Organizations have developed strategies to address these barriers, including partnering with faith-based institutions in Chicago (Kyle and Kearns 2018) and conducting civic ecology practices in ways that are nonthreatening to local and national governments, such as litter cleanups in cities in India and Pakistan (Abhyankar and Krasny 2018) and Iran (Kassam et al. 2018), and mangrove restoration in China (Abigail 2016). Community gardening may be particularly effective in helping individuals form an environmental identity, given that gardening allows one to express oneself through planting choices, feel competent in successfully growing plants, and contribute to one's community (Clayton 2012), and often entails cooperation among people from different walks of life (cf. Dovidio et al. 2007) and even engagement in a larger civic agriculture movement (cf. Polletta and Jasper 2001; Lyson 2004).

Environmental educators should consider factors that influence students' identities prior to those students' participation in an environmental education program, including family background, cultural norms, and positive and negative nature experiences through family and school. Participants' personal characteristics, such as openness to new knowledge and willingness to think critically, also affect their ability to form an environmental identity (Blatt 2013). When parents question their children's emerging environmental identities, environmental educators can create situations where parents will support their children's environmental identities, such as inviting parents to students' public presentations about their environmental projects (Stapleton 2015). In working with urban or other populations who may have had negative experiences in nature and limited opportunities for voluntarism, providing positive experiences in nature and opportunities to engage in environmental advocacy or stewardship, and to be recognized for these actions, are especially important to forming an environmental identity.

When people identify with polarized social groups, educators can foster cooperation by invoking a superordinate identity that all participants share (Fielding and Hornsey 2016). For example, the nonprofit organization Oldman Watershed Council in Alberta, Canada, "works with a variety of people and organizations, including motorized recreationists, campers, anglers and boaters," to engage them in restoring places where Albertans recreate, changing their behavior to reduce impacts and help them become better environmental stewards (OWC 2017). Whereas traditionally these groups have social identities that come into conflict (e.g., off-road recreationalist versus backpacker), the watershed council uses a superordinate identity garnered from in-depth interviews and conversations with diverse recreationists to help achieve its goals of people working together as land stewards. Regardless of whom they are trying to involve in stewardship, the council refers to everyone as "Good Albertans." The council also conducts activities that appeal to all recreationalists, such as volunteer hands-on river restoration days. This cooperation can further help build their shared identity as Good Albertans, which may promote future cooperation (Dovidio et al. 2007).

In short, regardless of the type of experience you provide, keep in mind that environmental identity:

- both leads to and is developed through participation in environmental action;
- develops through social interactions, with different interactions influencing different stages of identity formation; and
- is fostered by recognition by others (Stapleton 2015).

Assessing Environmental Identity

The Environmental Identity Scale measures the extent to which the natural environment plays an important role in how one defines oneself, and is focused on an individual rather than social environmental identity (Clayton 2003). It includes five factors: how salient the natural world is to one's identity (sample question: "I spend a lot of time in natural settings"); self-identification with nature (e.g., "I think of myself as a part of nature, not separate from it"); agreement with an environmental ideology (e.g., "Behaving responsibly toward the Earth—living a sustainable lifestyle—is part of my moral code"); positive emotions associated with nature (e.g., "I would rather live in a small room or house with a nice view than a bigger room or house with a view of other buildings"); and memories of spending time in nature (e.g., "I spent a lot of my childhood playing outside") (Clayton 2003, 45–46; see appendix). Other researchers have used open-ended interviews to gain an in-depth understanding of environmental identities, asking respondents to list up to twenty words or phrases answering the question "Who am I?"; to describe how their awareness of environmental damage originated and developed over time and what being an environmentalist means to them; and to list their environmental behaviors (Kitchell, Kempton, et al. 2000). Finally, art projects like constructing nicho boxes (cardboard boxes where children place pictures, symbols, figurines, writing, and other small objects that express their identity) can be used as embedded assessment (Derr et al. 2018).

12

NORMS

Highlights

- Social and personal norms are standards of behavior that society expects of us or that we expect of ourselves.
- Descriptive social norms refer to actual behaviors, injunctive social norms refer to behavioral expectations of one's group or society, and trending social norms refer to behaviors that are becoming more common.
- Personal norms are individuals' expectations for their own behaviors.
- Social norms influence environmental behaviors because people tend to conform with what others do.
- Personal norms influence environmental behaviors through feelings of moral obligation, guilt, and pride.
- Schools and environmental education programs and facilities can set social norms for behaviors and thereby influence students' personal norms and behaviors.

The motto of the litter cleanup organization the Ugly Indian is "See the change you want to be." It reflects that descriptive norms—what we *see* others doing, like littering or picking up trash—have a strong influence on what we do. The importance of descriptive norms is further emphasized in this passage from the Ugly Indian.

> We walk where more people are walking.
> We jump the signal if others do.

We remove footwear where others do.
Similarly we spit and litter where others have left the cues, and
We respect all the spaces that others have left untouched or clean.
(Abhyankar and Krasny 2018, 240)

In this chapter we explore sometimes surprising results from research on descriptive, injunctive, and trending social norms and personal norms.

What Are Norms?

People are less sensitive to what others feel they should do than what others actually do.

(Huber et al. 2018, 4)

Norms are standards of behavior that we expect for ourselves or that society expects of us. Humans have both social norms, which are related to group expectations, and personal norms, or expectations for ourselves as individuals (table 12.1).

Social norms come in three types. Social norms can be *descriptive*, in which case they describe what most people actually do. They can also be *injunctive*, related to behaviors we think are approved or disapproved by our social group or society. Descriptive norms tell us what everyone is doing, whereas injunctive norms tell us what "ought" to be done (Cialdini et al. 1990). Injunctive norms are also referred to as *perceived social* norms (Ajzen 1991). Recently, researchers have identified dynamic or trending norms to describe behaviors that

TABLE 12.1 Norms and related constructs

CONSTRUCT	DEFINITION
Personal norm	Expectations individuals set for themselves about how they act in particular situations (Schwartz 1977)
Social descriptive norm	What we believe other people do (Cialdini et al. 1990)
Social injunctive norm (also called perceived social norm)	What we believe we should do or behaviors we believe others will approve or disapprove of (Cialdini et al. 1990)
Social dynamic or trending norm	What we believe other people are doing more often (Mortensen et al. 2019; Sparkman and Walton 2017)
Choice architecture	Designing programs or facilities so that engaging in desired behaviors is easy or automatic (Sunstein and Reisch 2014)
Nudge	Aspect of a program or facility (choice architecture) that alters people's behavior without using force or economic incentives (Thaler and Sunstein 2008)

are changing or becoming more common (Mortensen et al. 2019; Sparkman and Walton 2017).

Personal norms are the expectations individuals set for themselves about how they act in particular situations (Schwartz 1977). They are often related to feelings of guilt or pride. If we act in ways that go against our personal norms, we feel guilty, whereas if we act consistently with our personal norms, we feel proud. In short, personal norms are about what we feel morally obliged to do (Bamberg and Möser 2007; Van der Werff and Steg 2015).

Why Are Norms Important?

Social norms and personal norms impact environmental behaviors. They work alongside other means of social influence, including regulations, and even the designing of buildings and programs, so that the "green choice" is the easy choice.

- Social and personal norms are important influences on individual behaviors and collective action.
- Descriptive and trending social norms appear to be particularly effective in influencing individual behaviors (Mortensen et al. 2019; Nolan et al. 2008).
- Social norms offer an alternative to government regulation or policies that use penalties to influence behavior. They may even be more effective than government policies, which sometimes promote resentment or backlash leading to noncompliance (Cialdini 2007).
- Social norms influence people's personal norms of responsibility and behavior (Bamberg and Möser 2007).
- Personal norms influence our individual behavioral choices, including our support for policies that entail personal sacrifices like paying more for green energy, and environmental citizenship behaviors such as joining an environmental group (Stern et al. 1999).

How Do Social Norms Influence Environmental Behaviors and Collective Action?

San Francisco passed a similar ban in 2014 on the sale of plastic water bottles 21 ounces or smaller in public spaces, including municipal buildings, streets and parks. David Chiu, a Democratic state legislator who championed the ban as president of the city's

board of supervisors, said it has helped change the culture. People now carry around their favorite refillable water bottle. Concerts and festivals set up refillable water stations instead of selling plastic bottles by the caseload. "While it was controversial when it was proposed, in hindsight today, it has been a no-brainer," he said. "I think everyone has adjusted to it."

(Hu 2018)

How do norms influence behaviors? Do different types of norms have different pathways to environmental behaviors and collective action?

Descriptive social norms influence behavior by providing information about what others are doing and thus what might be a "wise thing to do." Injunctive social norms can mobilize people to adopt a behavior by helping them evaluate what action will bring social approval or disapproval. Studies point to descriptive social norms as being more effective than injunctive social norms in influencing environmental behavior (Cialdini 2007; Farrow et al. 2017; Huber et al. 2018). Further, messages about descriptive norms are more effective than other types of environmental messages in influencing behaviors. In one study, residents who received a message with information about the conservation behavior of their neighbors were more likely to reduce energy use than those who received messages about how saving energy would save them money, help the environment, or benefit society. In fact, knowing what neighbors were doing was so influential that residents whose energy use was less than the average increased their energy consumption, presumably to be more like their neighbors (Nolan et al. 2008). The researchers termed this unfortunate outcome the "boomerang effect" and set about exploring how to avoid it. It turns out that when the residents who were energy saver "overperformers" also received a message rewarding their behaviors—a simple emoticon suggesting they were doing the right thing—they maintained their low energy consumption (Schultz et al. 2007). In a similar study of descriptive norms focused on what influences hotel guests to reuse towels, researchers also found that a message about a descriptive norm—that the majority of other guests at this hotel were reusing their towels—was more influential than a message about the importance of environmental protection. Descriptive norm messages are particularly effective when they reflect the behavior of groups with which participants identify or that closely match the situation and setting one finds oneself in (Goldstein et al. 2008). And when pro-environmental descriptive norms are accompanied by injunctive norms about what people ought to do, we can expect even greater compliance with environmental behaviors (Thøgersen 2008).

The boomerang effect wherein people alter their behavior toward the descriptive norm, regardless of whether that norm is pro-environmental, poses a problem. How can we influence people to engage in an environmental behavior if it's uncommon or "not the norm"? Here we can use dynamic or trending norm messages to influence behavior (Mortensen et al. 2019). For example, college students who were told that three in ten people had recently *shifted* to eating less meat were more likely to order a vegetarian meal than students who were told that three in ten people eat less meat (Sparkman and Walton 2017).

Personal norms reflecting a strong moral obligation also influence people's environmental choices. People may feel guilty if they behave in a manner not in accord with their personal norms and pride when they act consistently (Bamberg and Möser 2007). In one study, feeling "personally upset" when someone did not recycle—a measure of a personal recycling norm—predicted recycling behavior (Viscusi et al. 2011; Huber et al. 2018). Thus, through reflecting feelings related to specific behaviors, personal norms explain why people engage in those behaviors even when they may be costly in terms of one's comfort or pocketbook (Van der Werff et al. 2013).

To encourage environmental behaviors, researchers have experimented with means to "activate" norms or make them salient. Activating social norms makes people focus on them (Cialdini et al. 1990). Several factors can activate pro-environmental norms, including (1) awareness of the problems caused by our consumption behaviors, (2) acceptance of our personal responsibility for the problem, and (3) an expectation that one can help solve the problem by changing one's behavior (Steg and de Groot 2010; Van der Werff and Steg 2015).

Back in the 1980s, social scientists conducted an experiment to determine how descriptive norms and their salience influence littering behaviors (Cialdini et al. 1990). They used a parking garage as the setting for their experiment and exposed random passersby walking from the elevator to their car to one of four conditions. Each condition was meant to simulate different descriptive norms combined with different degrees of salience. A descriptive norm of "people do not litter" was simulated by a garage with no litter, whereas litter strewn around the garage simulated a descriptive norm of "people litter." To make the descriptive norm salient, a person who was part of the research team dropped a flyer in front of the passerby. For the norm-not-salient condition, the research team member simply walked by the passerby without littering (table 12.2).

Once the passersby walked through the garage and got to their car—having experienced one of the four treatments—they found a flyer the experimenters

TABLE 12.2 Norm littering experimental treatments and results (Cialdini et al. 1990)

NORM SALIENCE	DESCRIPTIVE NORM	
	NO LITTERING NORM	LITTERING NORM
SALIENT	Clean parking garage Person walks by and drops flyer in front of passerby Result: Passerby least likely to litter	Littered parking garage Person walks by and drops flyer in front of passerby Result: Passerby most likely to litter
NOT SALIENT	Clean parking garage Person just walks by	Littered parking garage Person just walks by

had placed on their windshield. A hidden observer then recorded whether the passersby littered the flyer or not.

It turns out that passersby once having returned to their cars were *least* likely to litter in a clean environment ("no littering" descriptive norm) and when the no-littering norm was made salient (the person dropped the flyer in front of the passerby). They were *most* likely to litter when they were in a littered environment and, similar to the situation where people did not litter, the other person littered in front of them, thus again making the norm salient. The researchers concluded that the descriptive norm of not littering was most important in influencing environmental behavior, but that activating or making a norm salient also played a role.

Visual signs indicating what other people do also influence environmental behaviors (Cialdini 2007; Nolan et al. 2008). For example, visitors to the Petrified Forest National Park in the United States were stealing highly valued petrified wood. To address the problem, social scientists worked with park officials to install two types of signs that would reflect different descriptive norms and environmental messages. One sign showed a picture of three people stealing wood, and another showed one person stealing wood; both signs were accompanied by a message urging visitors not to take wood. Visitors passing the sign with more people stealing wood were more likely to steal the petrified wood. This suggests that the signs' visual messages about descriptive norms—whether lots of people were stealing wood—impacted visitors' behavior (Cialdini 2003).

Social norms may be more important in influencing pro-environmental behaviors in collectivist cultures like China and Japan relative to individualistic cultures in the West (Clayton 2012; Eom et al. 2016). In fact, norms related to conformity and cooperation can also influence environmental behaviors and collective action. For example, employees may volunteer to plant trees for a corporate social responsibility day because they view cooperation with colleagues as the social norm for their workplace. Because social norms of cooperation are more widespread than environmental norms, they should be considered as a tool for fostering environmental behaviors (Kinzig et al. 2013).

Author Reflections

Littering happens to be an important issue for me—I think because it's so visible in my daily life and a reminder of larger environmental problems and attitudes. Each day I walk to work, I pass by litter on the grass next to the sidewalk. I live in Ithaca, New York, with excellent trash and recyclables collection, and with an expectation—that is, an injunctive norm—that people don't litter. Yet the evidence—the plastic cups, Styrofoam carry-out trays, candy wrappers, and beer cans—demonstrates that people do litter. Thus the descriptive norm seems to be in conflict with the injunctive norm. I also have a strong personal anti-littering norm—so strong that I daily pick up trash while walking to and from work. My personal norm becomes activated whenever I see a piece of litter. And I often feel guilty when I don't pick up a piece of trash. Because picking up trash is definitely not the descriptive or even injunctive social norm, I also feel a slight bit embarrassed when other people see me. When I mentioned my embarrassment to a colleague, he responded that I should feel proud, because by setting the example of picking up trash in public spaces, I would change the social norm, and more people would join me in cleaning up the trash.

To sum up, one can readily imagine how social and personal norms related to the environment might influence environmental behaviors. For example, if I see most people in the office walking up the stairs (descriptive social norm), and if I perceive that my colleagues think I should walk up the stairs rather than take the elevator to save energy (injunctive social norm), I am likely to walk up the stairs. I am also likely to walk up the stairs if I have a personal norm of saving energy and feel guilty if I take the elevator. Social norms influence behaviors through reflecting what other people do and what is the right thing to do. They also influence personal norms, which in turn influence behaviors through moral guidance. But to exert their influence, norms need to become activated or salient (figure 12.1).

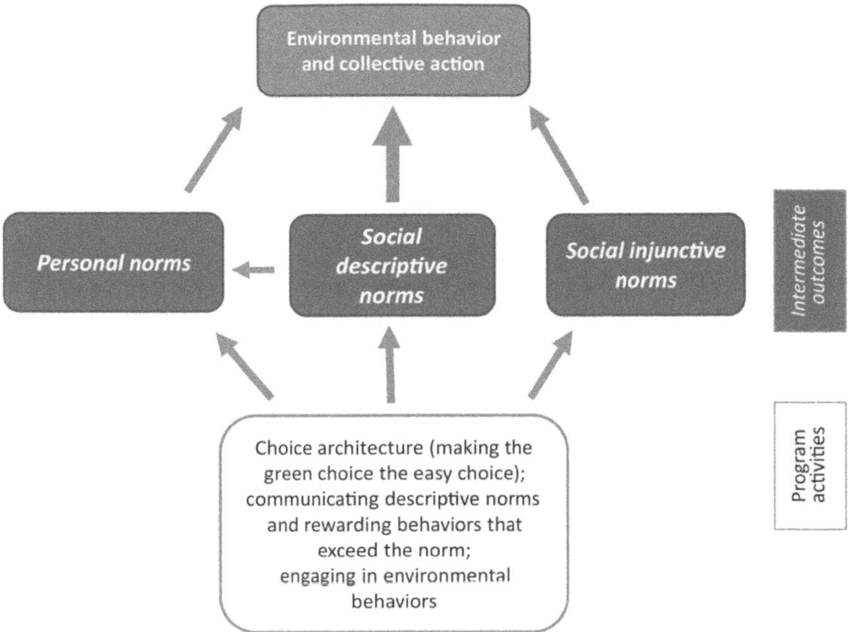

FIGURE 12.1 Personal norms influence behaviors through feelings of guilt or pride. Social descriptive norms have a stronger influence on environmental behaviors than social injunctive norms. Social norms influence personal norms.

How Can Environmental Education Nurture Social Norms?

> **If choosers know that most other choosers are selecting green energy, they will be more likely to choose green energy themselves. By contrast, if environmentalists lament the fact that few people are choosing green energy, they may aggravate the problem by drawing attention to, and thus reinforcing, a social norm that they hope to change.**
>
> (Sunstein and Reisch 2014, 129)

Environmental educators can foster pro-environmental norms using three strategies: (1) changing behaviors through design choices, (2) communicating existing norms in ways most likely to change behaviors, and (3) direct instruction to activate norms.

Using Design Choices to "Nudge" Behaviors

Let us assume that the participants in your program do not already hold pro-environmental norms. In this case, ask yourself: Are the behaviors that program participants see when they come to our program consistent with the pro-environmental norms I want them to adopt? For example, are educators using reusable coffee cups? When we host a meal, what foods are we serving, what kind of dishes are we using, and how are we reducing and handling waste? What kinds of toilets are at our center? Changing these "descriptive norms"—that is, the actual behavior students observe at a center, school, or other facility—impacts participants' behaviors. This influence is often stronger than other factors, including messages about environmental problems.

You might not be able to install composting toilets at your center, but likely you can make at least one or two choices that will determine the behaviors participants and visitors observe. In a simple example, pastors at churches in Chicago replaced plastic water bottles with glass cups on their lectern during sermons, thus setting an example and helping create norms for their congregation (Kyle and Kearns 2018). Similar to how many airports—and the building I work in at Cornell—have installed refillable water bottle stations, schools and environmental education centers can also make design choices that influence behaviors and set norms. A nature center can start using reusable dishes, and students can either rotate dishwashing responsibilities or wash their own dishes, similar to what I experienced at Hangzhou Botanic Gardens (see Author Reflections below). Or a community center might install a compost bin, and students might then transport the finished compost by wheelbarrow to a nearby garden.

By changing what options are available to program participants and center visitors, you are making the "automatic choice" the "green choice." If a cafeteria serves only vegetarian meals, not eating meat becomes the "automatically green" choice (Sunstein and Reisch 2014). Designing the environmental education center or other context to promote green choices eventually transforms social norms. Such design is referred to as "choice architecture" (Thaler and Sunstein 2008).

Author Reflections

Last spring, I ate lunch at the canteen in the Hangzhou Botanical Gardens in China. The canteen did not supply dishes, so we all had to bring dishes with us to get served lunch. Thus the default was to bring one's own dishes. And because the canteen had sinks for people to wash their own dishes, it was easy to bring reusable rather than throwaway dishes. The canteen's "choice architecture" led us to adopt the environmentally sound behavior, or the "automatically green" choice (Sunstein and Reisch 2014).

After a couple of days eating lunch at the Hangzhou Botanical Gardens canteen, we all got used to the system, and bringing and washing our own dishes became the descriptive social norm. Ideally, once we became accustomed to this social norm, we might adopt a personal norm of bringing our own reusable dishes to other settings like work or school. By allowing people to "see the change we want them to be," we might help spread this norm to additional settings (Ross et al. 2016; Huber et al. 2018).

Social norms and choice architecture "nudge" people to make environmentally sound choices without being coerced by government regulations or incentives (Thaler and Sunstein 2008). A nudge is "any aspect of the choice architecture that alters people's behavior in a predictable way" without forcing them or offering economic incentives (Thaler and Sunstein 2008, 6). An app that tracks sustainability and fitness behaviors and which is continually present on your phone lock screen and wallpaper is an example of a nudge. Given how often people check their phones, the display serves to "nudge" users to achieve their sustainability and fitness goals (Landay and Crum, n.d.).

Of course, a school or center can also establish firm rules—for example, straws are prohibited at our facility. Sometimes establishing rules leads to new social norms, such as government regulations on leaded gas that eventually become the norm for people at the gas pump (Kinzig et al. 2013). However, establishing rules does have drawbacks. For example, people may be more likely to react negatively to a written rule that they deem unfair than to less intrusive strategies to "nudge" behaviors. This may lead to resistance and people trying to get around the rule.

You might challenge your program participants to become "choice architects" for their nature center, community center, school, or home. They can explore what aspects of the setting influence staff, visitors, program participants, or family members to make particular choices. They can then design changes to foster "automatically green" choices—settings where the easiest behavior is the green behavior (Sunstein and Reisch 2014).

Communicating Social Norms

How you communicate social norms will depend on whether the norm is widely accepted among your audience. Whereas research has shown that descriptive norms are generally more effective than injunctive norms, in cases where the desired behavior is not the norm, you may need to use injunctive or trending norm messages.

Let's consider eating less meat and more vegetables, which is one of the top ten actions to reduce greenhouse gases (Drawdown, n.d.). Unfortunately, meat consumption has increased in the United States and globally in recent years, so you cannot use a descriptive norm that communicates "everyone is avoiding meat, so why not you?" In cases where the desired norm is not common practice, you can use injunctive norms that communicate about what people ought to do, as well as trending norms to communicate that a behavior is becoming more common. To increase your chances of influencing the target behavior, you may want to use a message that is specific about what not to do and, if appropriate, offers alternatives, such as lower your meat consumption to one or two meals each week and substitute nuts and grains for meat in those other meals. Do not send a message about how horrible it is that most people are increasing their meat consumption, as others may follow what most people are doing regardless of whether you describe the behavior as negative or positive (Cialdini 2003).

If the intended behavior reflects what people are actually doing, you can use a descriptive norm but with one important caveat. You will want to consider who among your participants already participates in the behavior. Some 25 percent of US residents volunteer an average of fifty-two hours per year (Bureau of Labor Statistics 2016), including in environmental stewardship—for example, in litter cleanups, oyster restoration, and neighborhood tree plantings (Krasny et al. 2014; Fisher et al. 2015; Krasny and Tidball 2015; Krasny 2018). For your program participants who don't already volunteer, you can communicate a message about how prevalent volunteering is, including among their peers, such as "over 26 percent of US teenagers volunteer 52 hours a year" (Bureau of Labor Statistics 2016). However, if you have program participants who volunteer *more* than the descriptive norm for their peers, they may reduce their volunteer activity to conform with the norm. In this case, include in your message "rewards" for their good behavior—anything from smiley faces to public recognition. In short, for the program participants whose volunteer activity falls below the descriptive norm, communicating the norm and its importance may encourage them to volunteer more often, whereas for those above the norm, add a message about how valuable their volunteering is (Farrow et al. 2017).

Direct Instruction to Activate Norms

Often environmental education programs emphasize environmental problems. Although this risks making participants feel helpless or even communicating a descriptive norm of "everyone is destroying the environment so why don't I join in?," research on norms suggests ways to make such instruction effective in fostering environmental behaviors.

Let's say your program wants to activate a social norm of avoiding single-use plastics. You will want to share three types of information to focus participant attention on the no-plastics norm, and thus to change their behaviors (Steg and de Groot 2010; Van der Werff and Steg 2015).

First you want participants to be aware of and knowledgeable about the plastics problem (problem awareness). For example, in trying to change choices about plastics use, an environmental educator might include messages like "eight million metric tons of plastic ends up in the ocean every year" (Ocean Conservancy 2018), or show a video of an endangered sea turtle with a plastic straw stuck in its nose (Cuda and Glazner 2015). But we know negative environmental messages alone are not so effective in changing behaviors.

Coupled with the message about the environmental problem, you want your program participants to be aware of their responsibility in creating the problem. Reflecting how straws contribute to the global plastics problem, Diana Lofflin, founder of StrawFree.org, tries to create awareness of the problem and one's personal responsibility by telling people, "You use a straw for 10 minutes, and it never goes away" (Morgan 2018). This step is referred to as acceptance of personal responsibility.

Finally, and importantly, you want to show program participants what they can *do* about the problem. Often environmental problems seem overwhelming, and participants feel as if their behaviors will never make a difference. In this case, you can break down the big problem into smaller components over which participants have control. For example, plastics are a seemingly insurmountable global problem, but plastic straws are an important component of the problem and are easy to stop using. This last step is referred to as communicating outcome efficacy, and often engages participants in the desired behaviors. It is similar to action-related knowledge (see chapter 6) and reflects what we know about the importance of self- and collective efficacy in influencing environmental behaviors and collective actions (see chapter 10).

In short, use the following guidelines for using norms to change behavior:

- Make the green choice easy by changing the "choice architecture." You can do this by changing policies or infrastructure (e.g., eliminate plastic cups and cutlery at your programs, serve meatless meals).
- Create environmental descriptive norms at your facility or school so people "see the change" you want them to make.
- Use descriptive norm messages rather than messages about environmental problems.

- Use messages that emphasize how many people are engaged in the environmental behavior rather than the prevalence of negative behaviors.
- Reward people who exceed the norm.
- Use "ought" messages only where there is no positive descriptive norm and most people are below the norm.
- Engage your participants in learning about environmental problems, their personal responsibility for the problem, and what they can do to address the problem.

Assessing Norms

You might want to know whether your attempts to change the social norms at your facility or program are working. You might also ask, What are the personal environmental norms of the participants in my program? And is my program having an impact on participants' social or personal norms?

Social norms can be assessed by factual data on a city-wide or other basis. For example, an abundance of solar panels on homes in a city would be an indicator of a social norm of using clean energy. One might also measure things like declining use of plastic water bottles or straws as an indicator of changing social norms at a school or community center.

You could also measure your participants' injunctive social norms. So, for example, you could ask participants whether they think a fellow program participant would be upset if someone littered, if someone drove a gas-guzzling car, or if their neighbor watered their lawn during a drought. For personal norms, you can ask participants about their feelings of guilt or pride related to environmental behaviors (Van der Werff et al. 2013; see appendix).

In sum, you can assess whether any attempts at "choice architecture"—that is, changing your facility or program to make green choices the easy or automatic choice—are actually influencing your school or center's use of plastics, composting, or other practice—an indicator of descriptive social norms. You can also assess whether the changes you are making to your facility or program are having an impact on how participants perceive the social norm in your program, as well as on participants' feelings of moral responsibility to engage in environmental behaviors.

13

SOCIAL CAPITAL

Highlights

- Social capital encompasses relationships built on trust and norms that facilitate cooperation for mutual benefit.
- Social capital fosters collective action to support a public good such as environmental protection, and in some cases is associated with environmental behaviors.
- Social capital facilitates collective action by reducing transaction costs and creating opportunities for exchange of information.
- Environmental education can foster social capital by engaging participants in challenging, cooperative activities, offering the support needed for the group to succeed, setting norms of fairness and open communication, and providing opportunities for participants to partner with community members.

Let's say you want people in your community to lend their support to a public good, like a city park, bike lanes, or replanting a village forest. You are thinking about educating people on why these things are important, but you doubt that such an effort will make a difference. Over a hundred years ago, Lyda Hanifan, the state supervisor of rural schools in West Virginia, faced a similar problem. His state suffered from poor roads, and he saw the solution as getting rural people to advocate for road improvements. Rather than lecture people about the importance of roads as a public good, Hanifan decided to invite them to "'sociables,' picnics and a variety of community gatherings," which would build trust and social connections. He bet that "then by skilful leadership this social capital may easily be directed towards the general improvement of the community wellbeing" (Hanifan 1916, 131). Hanifan is thought to be the first person to have used the term "social capital."

Hanifan's approach was a success. The social activities enabled rural people to accumulate social capital, which they leveraged to successfully advocate for road improvements. Reflecting later ideas about people as agents who both create and benefit from social capital (Portes 1998), Hanifan noted, "The more people do for themselves the larger will community social capital become, and the greater will be the dividends upon the social investment" (Hanifan 1916, 138).

What Is Social Capital?

Social capital refers to trusting relationships among members of a community that confer direct benefits on individuals and communities. Social capital can be broken down into three components. First is social connections with others garnered through engagement in civic life and voluntary associations. Specific trust for people you know or generalized trust in others is the second component. The third aspect is shared social norms, or formal and informal rules that guide our behaviors (Ahn and Ostrom 2003, 2008; see also chapter 12).

Over the years, social capital has assumed different meanings, varying in their emphasis on community, individual, and society-wide benefits (table 13.1; see also Claridge 2004). Hanifan (1916) used the term to describe how as people engaged in community social events and got to know and trust each other, they became advocates for public goods like better roads. Later, the term was used to describe the ability of individuals to secure benefits, like a job or admission to an elite college, through their social networks (Bourdieu 1986; Coleman 1988). Harvard political scientist Robert Putnam (1995) directed attention to the work of voluntary associations, including those focused on environmental stewardship, in his definition of social capital (Sirianni and Friedland 2005; Klyza et al. 2006; Krasny and Tidball 2015). Putnam and others also distinguished between bonding social capital among homogeneous individuals within groups and bridging social capital across people from different groups (Coffé and Geys 2007).

Whereas Lyda Hanifan focused on roads, Nobel laureate Elinor Ostrom saw social capital as a prerequisite for communities working together to sustainably manage natural resources held in common, such as irrigation water or forests (Ahn and Ostrom 2003). She wanted to understand how communities overcome social

TABLE 13.1 Social capital definitions

AUTHOR	DEFINITION	EXPLANATION
Hanifan (1916, 130, 131)	"Goodwill, fellowship, mutual sympathy and social intercourse among a group of individuals and families who make up a social unit," which can be "directed towards the general improvement of the community well-being"	First use of the term "social capital"; emphasizes building social connections to motivate rural Americans to advocate for a public good
Bourdieu (1986, 248)	"The aggregate of the actual or potential resources which are linked to . . . membership in a group"	Emphasizes individual benefits that come through social connections
Coleman (1988, S98)	"A variety of different entities having two characteristics in common: They all consist of some aspect of social structures, and they facilitate certain actions of actors . . . within the structure"	Emphasizes the function of social capital in close-knit groups as reinforcing positive social norms and thus conferring advantages on the group
Putnam (1995, 67)	"Features of social organization such as networks, norms, and social trust, that facilitate coordination and cooperation for mutual benefit." Can be bonding among people within a group or bridging between people in different groups.	Drawing from democratic traditions dating back to de Tocqueville's work in the mid-1800s, emphasizes volunteer engagement in civic life, including clubs (e.g., Rotary), social recreational activities (e.g., bowling), and social welfare organizations (e.g., soup kitchens)
Ahn and Ostrom (2008, 73)	"A set of prescriptions, values, and relationships created by individuals in the past that can be drawn on in the present and future to facilitate overcoming social dilemmas"	Used in explaining how communities cooperate to manage common resources sustainably and to avoid negative outcomes like the tragedy of the commons

dilemmas, or situations where each individual benefits from looking after his own interests, but the community as a whole suffers. For example, a social dilemma or "tragedy of the commons" occurs when any one fisherman catches as many fish as he can, which initially benefits him and his family. However, because the other fishermen are also exploiting the same lake, eventually overfishing leads to environmental degradation and threatens all the fishermen's livelihoods. Whereas external controls, such as regulations, incentives, and privatizing land, are one means to avert the tragedy of the commons, Ostrom and colleagues (2008) noted that when social capital is present, such top-down approaches are not needed for people to manage common resources sustainably.

Why Is Social Capital Important?

> **Unlike fear-based approaches to sustainability, where people comply because they fear potential reprimands or financial penalties, the social capital approach focuses on developing a genuine commitment to sustainable behaviors as the norm.**
>
> (Miller and Buys 2008, 246)

Communities with social capital are more willing to manage commonly held environmental resources, such as forests or water, sustainably and for the public good (Dietz et al. 2003; Ahn and Ostrom 2008). Social capital also has important implications for environmental education and youth development.

- Social capital offers a framework for environmental education that shifts the focus from changing individual behaviors to creating the conditions that enable a community to take collective action to safeguard its commonly held resources (Krasny, Kalbacker, et al. 2013).
- Social capital offers a framework for environmental education to play a role in ever more frequent and devastating disasters and other types of change (Krasny et al. 2011; DuBois et al. 2017). This is because communities with social capital are more resilient—that is, able to adapt and transform in the face of ongoing change and large disasters (Folke et al. 2002; Walker and Salt 2006; Plummer and FitzGibbon 2007; see also conclusion).
- Social capital has societal and youth development benefits (see chapter 14). Societies with higher levels of social capital tend to have lower crime rates, better performing civic institutions (Lochner et al. 1999), and better school performance (Coleman 1988). For youth, social capital is associated with reduced rates of teen pregnancy and delinquency, enhanced happiness and health, high-quality relationships with adults, and civil society skills such as the ability to run meetings (Bettertogether 2000; Lewis-Charp et al. 2003; Jarrett et al. 2005; Ferguson 2006; Helve and Bynner 2007; Rossteutscher 2008).

Author Reflections

After the *Global Warming of 1.5°C* report (IPCC 2018) came out, I started thinking more and more about the need for collective action to address climate change. And I puzzled about the role of individual behaviors, which seem insignificant relative to the scale of the problem. I turned to social capital—the social connections each of us has with people we trust and who share our environmental values. In part owing to social media, we may share social capital with a relatively large group. And our social networks function at least in part like a collective or organization. Could this be a pathway to translate individual behavior change to more impactful collective action?

What if each of us mobilized our social capital to instigate collective action to reduce greenhouse gases? Consider your social network—the people you are connected to via Facebook, WeChat, Instagram, or another social media—as a collective whose members you can influence. Next consider a behavior you want to take to reduce your carbon footprint—say, reducing your red meat and dairy consumption. In February 2019, I launched the Cornell Climate Online Fellows program, through which thirty-five environmental educators from twenty-six countries are exploring how we can each leverage our social capital and influence our social networks to effect larger changes, going beyond simply acting as individuals (Krasny 2019).

How Does Social Capital Contribute to Collective Action and Environmental Behaviors?

Social capital facilitates collective action by lowering the transaction costs and the risks of working together. I am more likely to spend time going to a meeting to advocate for a community solar farm if I trust that my neighbors will also show up. If community gardeners trust that their fellow gardeners are putting in their fair share of time weeding, they will be more likely to keep their own plots tidy and volunteer to weed the garden pathways and fix the garden fence. Importantly, people who trust each other don't need to spend time monitoring others to make sure everyone is contributing a fair share of work; rather they can simply engage in collective action for the public good (Dietz et al. 2003; Pretty 2003). Such collective action also builds social capital (figure 13.1).

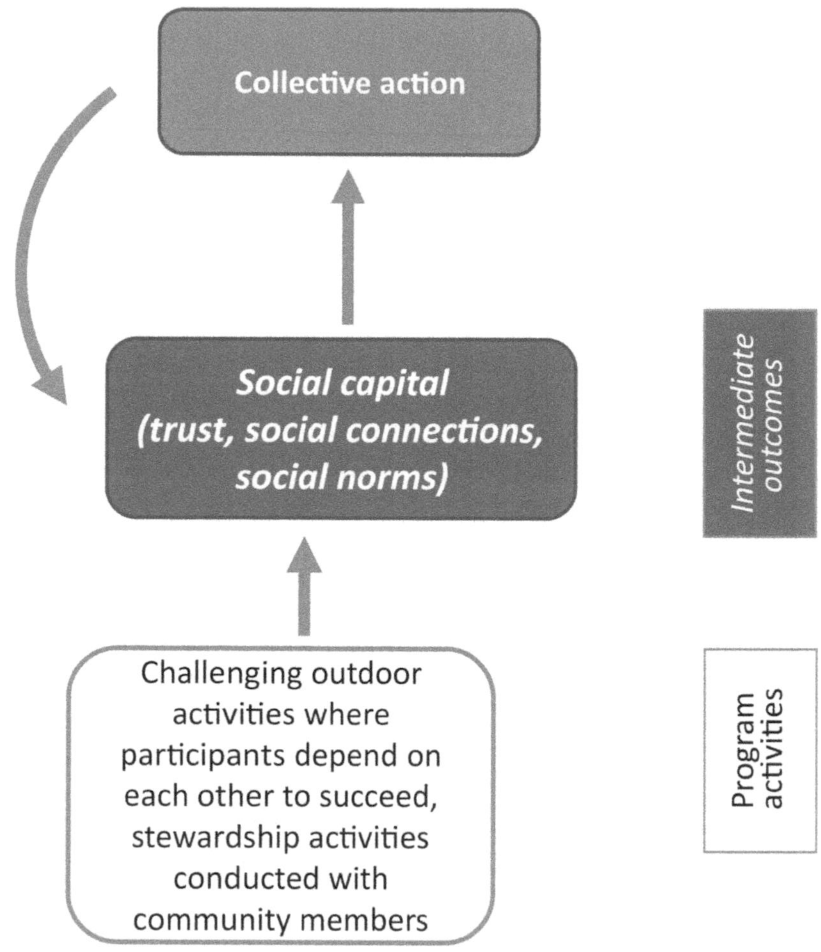

FIGURE 13.1. Social capital feedback pathway. Social capital leads to collective action, which also builds social capital.

Social capital can also foster environmental behaviors. Studies have shown that individuals with more social connections with neighbors, and those with strong connections with family members who practice green behaviors, are more likely to themselves practice environmental behaviors. This may be due to people learning about and seeing locally relevant examples of environmental behaviors from neighbors and family members, such as energy-saving tips and environmental groups they can join. People with strong connections also may become aware of ways to reduce the costs of environmental behaviors (e.g., learn about opportunities to carpool or join a solar collective), or feel pressure to conform with their neighbors' or families' green social norms (Miller and Buys 2008; Videras et al. 2012; Macias and Williams 2016; Cho and Kang 2017).

How Can Environmental Education Nurture Social Capital?

Environmental education can build social capital through fostering trust and social connections, and by setting norms of cooperation (see chapter 12). Strategies for building both in-group trust, as might occur when children learn to trust the other children and adults in their environmental program, and generalized trust, which refers to trusting people in society more abstractly, can be included in programs. Similarly, environmental education can build social connections within and outside a program. Trust and social connections within the program would be referred to as bonding social capital, whereas trust and social connections with other organizations are part of bridging social capital. Within-program or bonding social capital facilitates collective action among participants, whereas bridging social capital that includes people outside an environmental education program can enable students to engage in larger-scale collective actions (Krasny, Svendsen, et al. 2017; Stern and Hellquist 2017).

Several environmental education strategies can be used to build in-group trust among participants (McKenzie 2003; Ardoin, DiGiano, O'Connor, and Podkul 2016). In a residential program lasting from two and a half to five days, ten- and eleven-year-olds attributed the development of trusting relationships to being free from prescribed classroom roles, and to the program providing a safe space to take on different roles and to be supported in taking on new challenges. As they conquered physical challenges freed from classroom constraints and expectations about classroom learning, participants began to interact and offer support to peers, including students from their school with whom they had not previously interacted (Ardoin, DiGiano, O'Connor, and Podkul 2016). Succeeding in outdoor challenges, such as rock climbing and blindfolded "trust walks," depends on cooperation with fellow participants and can similarly build trust.

Environmental action programs also create trusting relationships among participants, and between participants and program leaders through cooperative learning and team building as well as fun activities like going out for ice cream. Also important are open and honest communications and respect for multiple viewpoints during discussions about what actions to take. In addition, educators wishing to build trust should demonstrate respect and caring for participants' lives outside the actual program, while providing scaffolding for leadership and other challenging roles (Schusler and Krasny 2010; Stern and Hellquist 2017; Delia and Krasny 2018).

Environmental education programs are often conducted in cooperation with civil society organizations—like friends of parks groups, nature centers, YMCAs, 4-H programs, and environmental monitoring and stewardship organizations. In places where many civil society organizations exist, where they are perceived as being fair and effective, and where they have committed and engaged volunteers cooperating for the common good, generalized trust and social connections, or bridging social capital, are stronger. Thus, strategies to build bridging social capital include actively engaging participants in service learning, intergenerational activities, and voluntary associations conducting watershed monitoring, citizen science, tree planting, advocacy for renewable energy, and similar activities that involve multiple organizations (Ballantyne et al. 2006; Wollebæk and Selle 2007; Schusler and Krasny 2010; Thornton and Leahy 2012; Lindberg and Farkas 2016; Van Deth et al. 2016; Delia and Krasny 2018). In Poland, where citizens generally have low levels of trust, in part due to having lived under communist rule, secondary school students built trust with community members through interviewing elders and historians to learn about their villages' multiethnic history and its past discrimination against Jews (Stefaniak et al. 2017). Finally, when parents come to their children's final presentations or other events, they may also form social ties (Offer and Schneider 2007).

In short, to enhance trust among your participants, engage them in challenging cooperative activities, while setting norms of fairness and open communication and offering the scaffolding and support needed for the group to succeed. You may also want to include fun social activities to build social connections among youth in your group. To help students form trust and connections outside their group, or bridging social capital, provide meaningful ways to engage with the community, through service learning, intergenerational learning, and partnering with other environmental, community, and business groups.

Assessing Social Capital

Because social capital encompasses multiple dimensions, including in-group and generalized trust and social connections within and outside an environmental education program, you will want to specify your social-capital-related goals in planning your assessment. Does your program aim to increase levels of trust among participants, between participants and program leaders, and/or between participants and community members? Are you concerned with students developing social ties where they share information and reinforce social norms? Are you interested in whether your program has motivated students to join civic groups in the future?

A scale designed to measure in-group trust among students before and after a program included items such as "How often do program participants keep promises?" "If you had a problem, would you go to a participant in this program?" and "If you had a problem, would you go to a leader in this program?" (Ardoin, DiGiano, O'Connor, and Podkul 2016). Generalized trust can be measured by items such as "I trust most people in my neighborhood," or "In general, I can trust most people" (Krasny, Kalbacker, et al. 2013; see appendix).

To determine the strength of ties formed among students in your program, you can ask questions such as, "How often do you spend time with students in the program?" "How often do you communicate on social media with students in the program?" or "How well do you know students in the program?" If you were interested in knowing if your program increased bridging social capital, you might ask about how often program participants exchanged information with community members they met through the program.

Finally, an environmental education program might motivate students to expand their participation in voluntary associations. Here you might ask questions about what other volunteer, service learning, advocacy, or policy activities students have engaged in (Roper Center for Public Opinion Research 2000). If your program is short, students may not have an opportunity to increase these activities during your program. In this case, you might ask about students' intentions to engage in these activities in the future; however, intending is not the same as actually doing something, so the results should be reported as intentions rather than actual changes in behavior. Alternatively, you can conduct surveys several months after a program has ended.

14

POSITIVE YOUTH DEVELOPMENT

Highlights

- Positive youth development entails designing programs to foster social, emotional, intellectual, and physical well-being among youth.
- Positive youth development approaches to environmental education help youth acquire assets important to their success and to their ability to engage in environmental behaviors and collective action.
- Self-efficacy, bonding with others, trust, and civic participation are assets that enable youth to contribute to environmental goals.
- Environmental education programs foster youth assets by providing youth with a sense of belonging, challenges that lead to new skills and ways of thinking, and opportunities to have their voice heard and to make meaningful contributions.

Whereas politicians often pit environmental against social concerns, in fact environmental education can contribute to positive youth development. Often community centers in low-income neighborhoods see environmental activities, such as community gardening or monitoring water quality, as a means for youth to develop job, communication, civic participation, and other skills critical to their future. By partnering with youth development professionals, environmental educators can expand their networks, audiences, and programs, particularly in low-income and ethnic minority communities.

What Is Positive Youth Development?

> **If positive development rests on mutually beneficial relations between the adolescent and his/her ecology, then thriving youth should be positively engaged with and act to enhance their world.**
> (Lerner et al. 2011, 6)

Starting in the 1990s, interventions to support families and children took a turn away from focusing on problem behaviors of troubled teens. Instead scholars turned their attention to what factors are present when youth experience healthy physical, intellectual, emotional, and social development (Eccles and Gootman 2002; Catalano et al. 2004; Lerner and Lerner 2011). An outcome of this work is an asset-based approach, referred to as positive youth development, which assumes that all youth have the capacity to become successful adults, given appropriate support (Eccles and Gootman 2002; Lerner et al. 2005).

Developing youth to their full capacity as human beings entails attention to the interactions of youth with their social and physical environment. Thus, positive youth development scholars and practitioners consider both youth assets and the characteristics of settings that enable youth to develop those assets (Eccles and Gootman 2002). Assets include self-efficacy, pro-social norms, and meaningful relationships with peers and adults, as well as, more broadly, social, emotional, cognitive, behavioral, and moral competence (Catalano et al. 2004). One approach to positive youth development focuses on the "Six Cs," defined as competence, confidence, connection, character, caring, and contributions to community and civil society (Lerner et al. 2005).

Positive youth development emphasizes the two-way interactions between youth and their contexts. As youth develop the ability to contribute to their community and environment, they change the context in which they and other youth are able to realize assets. This is referred to as individual ↔ context relations. The feedback between youth and their surroundings is generally thought to be mutually beneficial (Lerner et al. 2011). For example, in an urban agriculture internship program in Brooklyn, New York, environmental educators provided a social context for youth interns to develop assets including responsibility, social connections, and leadership. In turn, experienced interns created a social context emphasizing belonging and acceptance for new interns. The interns also worked at an urban farm and farmers' market, as well as advocated to preserve community gardens, thus improving the physical environment for other youth and community members. Such social and physical improvements become part of the context for future youth development (Delia and Krasny 2018).

Six C's of positive youth development (Lerner and Lerner 2011)

Competence

Social: interpersonal skills including communication, assertiveness, conflict resolution
Cognitive: critical thinking, problem solving, decision making, planning, and goal setting
Academic: school achievement, attendance, graduation rates
Vocational: work habits and career choice explorations

Confidence: self-esteem, self-efficacy, identity, belief in the future
Connections: building and strengthening relationships with other people and institutions such as school
Character: decreasing engagement in health-compromising (problem) behaviors, respect for cultural or societal rules and standards, a sense of right and wrong (morality)
Caring: empathy and identification with others (in environmental education, this would include nonhuman life)
Contribution: to one's community through civic engagement

Social justice youth development takes into account structural inequities and barriers to youth acquiring assets, such as poverty, racism, sexism, homophobia, violence, and drugs (Ginwright and Cammarota 2002; Sukarieh and Tannock 2011), including in developing countries with severe gender inequity (Briggs et al. 2019). It seeks to cultivate among youth an awareness of social justice, the ability to respond to oppressive forces, an understanding of the root causes of social problems, and the ability to take social action that addresses larger political forces (Ginwright and Cammarota 2002). Social justice youth development draws on Freire's (1970) notions of critical consciousness (*conscientização*), where learners question their historical and social context and related injustices, and of praxis, or reflection and action in order to transform the world. Critical pedagogy of place is an environmental education approach that similarly engages issues of power and justice (Bowers 2002; Gruenewald 2003; see also chapter 9). The urban agriculture intern program in Brooklyn, New York, in which students examined structural issues impacting access to healthy foods, illustrates social justice youth development and critical pedagogy-of-place approaches (Delia and Krasny 2018).

Why Is Positive Youth Development Important?

A positive youth development approach enables environmental education to partner with organizations that address social concerns of a local community. By working together, environmental and community groups demonstrate that environmental issues need not be pitted against social issues.

- Positive youth development programs attract youth who may not have strong environmental interests and thus would be unlikely to participate in a more traditional environmental education program (Stephens 2015).
- In communities where youth face poor schools, poverty, and other threats to their development, environmental activities like community gardening become part of efforts to help youth succeed in school; form positive relationships with adults, peers, and family members; and develop communication skills, feelings of self-worth, social commitment, and responsibility (SEER 1985; Lieberman and Hoody 1998; Schusler and Krasny 2008, 2010, 2014; Riemer et al. 2014; Schusler 2014; Stephens 2015; Delia and Krasny 2018).
- Positive youth development contributes to environmental education goals of changing individual behaviors and collective action (Schusler and Krasny 2010; Schusler 2014). Once young people have developed communication skills, positive relationships with others, and a commitment to helping their community, they will be more likely to engage in other positive behaviors and collective actions, including those that integrate environmental and social outcomes (Flanagan and Levine 2010).
- The ability to address youth development goals enables environmental education programs to partner with youth and community development organizations, thus broadening environmental education's influence and outcomes (Fraser et al. 2015; Krasny, Chang, et al. 2017; Krasny, Danter, et al. 2017).

How Does Positive Youth Development Contribute to Environmental Behaviors and Collective Action?

Youth with social, intellectual, and emotional assets are more likely to engage in environmental behaviors and collective action (figure 14.1). For example, self-efficacy (or locus of control) is both a developmental asset and a predictor of environmental behaviors (Hungerford and Volk 1990; Chawla 2009;

see also chapter 10), and critical-thinking and decision-making skills are needed to solve environmental problems (see chapter 6). Additionally, positive youth development programs provide volunteer, service learning, and other opportunities to develop trust and social connections, or social capital, which enables collective action to conserve commonly held resources such as forests or urban parks (Ahn and Ostrom 2008; see also chapter 13). Finally, youth who are engaged in civil society are more likely to continue that engagement as adults (Flanagan and Levine 2010). As a leader of an environmental action program explained,

> I think it's very important that [youth] have the opportunity to learn, to have the experience of giving back to the community because if they have a positive experience as seniors in high school, they'll be more likely to be lifelong stewards, giving back to the community in some way. (Quoted in Schusler et al. 2009, 117)

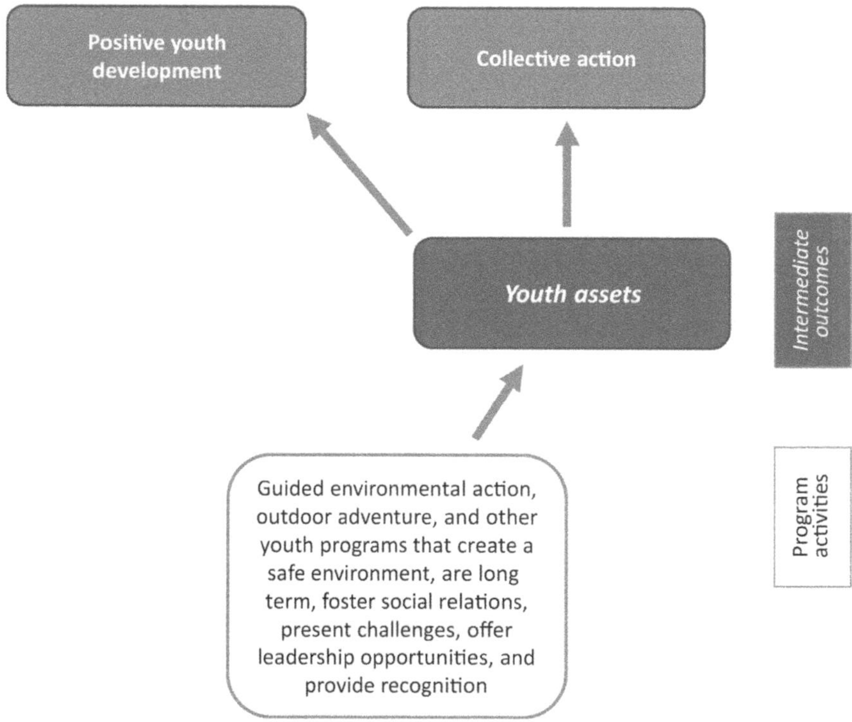

FIGURE 14.1 Youth assets like self-efficacy and communication skills enable youth to engage in community action. Positive youth development may also be a desired outcome for community environmental education.

Author Reflections

My undergraduate major was in adolescent development, and I volunteered as a tutor for youth living in detention facilities. After completing graduate degrees in forest ecology and becoming a professor at Cornell University, I came to see how youth development professionals were integrating environmental stewardship into their programs. I also came to appreciate the opportunities programs such as Rocking the Boat in the Bronx and East New York Farms! in Brooklyn provide for developing youth assets through collective environmental action. My friend and colleague Akiima Price calls these practices community environmental education, the goal of which is to enhance a community's wellness through thoughtful environmental action. Community environmental education fosters collaborative learning and action, taking into account the social, cultural, economic, and environmental conditions of a community (Price et al. 2014). For me, the beauty of community environmental education is that it recognizes the environmental work already going on in low-income communities and communities of color—such as community gardening, urban farming, neighborhood cleanups, and linking restorative justice for criminal offenders with environmental justice—and connects with and supports these existing efforts. Generally, the leaders of these initiatives are focused first and foremost on youth development, and use environmental stewardship, advocacy, and other environmental activities as a pathway for youth to build developmental assets.

How Can Environmental Education Nurture Positive Youth Development?

Environmental action programs (Schusler and Krasny 2008, 2010; Schusler et al. 2017; Delia and Krasny 2018), residential and outdoor adventure programs (McKenzie 2003; Stern et al. 2010; D'Amato and Krasny 2011; Powell et al. 2011; Ardoin et al. 2015; Ardoin, DiGiano, O'Connor, and Holthuis 2016; Williams and Chawla 2016), youth programs conducted by community organizations such as the YMCA and Boys and Girls Clubs (Larson 2000; Larson and Angus 2011; Lerner and Lerner 2011; Salusky et al. 2014), environmental clubs (Johnson et al. 2007; Johnson-Pynn and Johnson 2010; de Vreede et al. 2014; Comber 2016), and private-sector internships integrated with high school classes (Stephens 2015) all offer opportunities for youth to acquire developmental assets. Youth programs that foster positive youth development have several elements in common:

- They are long-term in duration and integrated with family, school, and community efforts.
- They provide physical and psychological safety and a sense of belonging.
- They foster positive and sustained relationships among youth and between youth and adults.
- They include activities that build life skills through setting expectations, posing challenges, and providing recognition.
- They empower youth by providing increasingly challenging opportunities for them to use these life skills as both participants in and as leaders of valued community activities.

(Roth and Brooks-Gunn 2003; Lerner and Lerner 2011; de Vreede et al. 2014; Schusler 2014; Stephens 2015)

Below we focus on how environmental action and outdoor adventure programs foster positive youth development.

Environmental Action Programs

It wasn't dictated by me and it wasn't just created by them either.

(environmental educator, quoted in Schusler et al. 2017, 14)

Environmental action programs follow "a participatory pedagogy in which learners analyze the causes of environmental problems and take action with others to generate and implement solutions" (Schusler et al. 2017, 533; see also chapter 5). Through voluntarism, service learning, and related forms of civic participation, youth develop assets (Lerner and Lerner 2011) and connect with their community. They also engage in critical reflection, sometimes focusing on issues of power and injustice (Schusler and Krasny 2010; Delia and Krasny 2018). One educator leading an environmental action program described her work as "preparing youth for future roles as voters who think critically about issues" and as "agents of social change within their communities" (Schusler et al. 2009).

Although youth are often painted as autonomous leaders in environmental action programs, educators play a critical role as they negotiate the "autonomy-authority duality of shared decision-making." Adult authority stems from decision-making power as well as from experience and wisdom. Approaches to balancing youth autonomy with adult authority described by educators include "structuring youth participation, supporting youth, valuing mutual learning, and communicating transparently to develop equitable partnerships" (Schusler et al. 2017, 533).

- *Structuring youth participation*. Educators structure youth participation to reflect learning goals, organizational and community contexts, and participants' existing assets. This entails creating a structured process for youth decision making by setting overall program goals while allowing youth to decide routes to achieve them. Educators also help youth envision and weigh the pros and cons of various options to achieve program goals. Example strategies range from allowing youth to decide what project they want to undertake and how to implement it; to the educator deciding the project focus and youth and adults sharing decision making about implementation; to the educator directing classroom activities while allowing youth to direct related activities in an after-school program.
- *Support*. Educators provide the support necessary for youth to succeed by conducting advance training for youth and scaffolding activities, asking guiding questions, facilitating reflection on individual performance and the group's collective progress, and helping youth to resolve conflicts.
- *Mutual learning*. Educators recognize themselves as learners alongside youth and value youths' experience and knowledge. One educator described this as "pilgrimage teaching," explaining that "I can either tell you, or I can show you, or I can go with you. And all . . . of us go with our students. We don't have the answers, but we have the energy to go with them and learn with them" (Schusler et al. 2017, 544).
- *Transparent communication*. Educators voice their opinions and explain why they make certain decisions, while allowing youth a safe space to express their own perspectives (Schusler et al. 2017).

The roles of youth and adults may change over the course of a program as youth take on greater responsibility and leadership. In the East New York Farms! urban agriculture internship program, adult leaders initially provide youth with a sense of belonging, but increasingly push them to take on ever more challenging tasks and to grapple with complex concepts related to plant growth, composting, and food justice. As the interns return in subsequent years, they take on greater program responsibilities, such as providing workshops for younger interns and community members, mentoring younger interns in their farm and farmers' market tasks, and participating in protests to preserve community gardens (Delia and Krasny 2018). Educators' work with youth demonstrates "authentic care," or making young people feel safe while challenging them to reach beyond what they think they can accomplish, or what society is telling them they can accomplish (Valenzuela 1999). The urban agriculture intern program also illustrates a social justice approach, as youth develop critical consciousness through posing questions about food security in their low-income community of color, which enables them to perceive, understand, and potentially counter oppressive food systems (Freire 1973; Delia and Krasny 2018).

Outdoor Adventure Programs

In outdoor adventure programs lasting several weeks or months, participants develop assets through the challenges, intensity, and novelty of the experience, and through opportunities to spend time in nature and be part of a tight-knit community (McKenzie 2000; D'Amato and Krasny 2011). Participants often describe these programs as transformational, which can be attributed to experiencing a "disorienting dilemma" (Mezirow 2000) stemming from living in the wilderness and adopting a lifestyle radically different from everyday life. To help participants process these dilemmas and acquire assets, instructors maintain a supportive social community, set expectations, provide opportunities to master the physical and social challenges inherent to such programs, and set time for self-reflection to process feelings of disorientation (McKenzie 2000, McKenzie 2003, D'Amato and Krasny 2011). These strategies may be adapted for shorter-term, nearby outdoor adventure or challenge experiences.

In short, environmental educators seeking to foster positive youth development should provide youth with a sense of belonging and with positive social connections, while pushing them to take on increasing responsibility and leadership. As educators navigate providing guidance and allowing youth to realize their potential, they constantly reflect on their own role and are open with youth about their reflections. They are also willing to adapt their approach as youth acquire newfound abilities.

Assessing Positive Youth Development

Surveys are available for measuring youth assets in developed and developing countries. In addition to asking youth about their perceived self-efficacy (see chapter 10), trust (chapter 13), and well-being (chapter 15), surveys include questions about youth engagement in advocacy, mentoring younger youth, clubs and after-school programs, and volunteering (Lerner et al. 2005; Zaff et al. 2010; Hinson et al. 2016; see appendix). Surveys of parents' perceptions of their children can also be used to assess changes in youth assets and civic engagement.

Open-ended and narrative interviews and participant observation conducted using an appreciative inquiry process provide in-depth understanding of how youth acquire assets. For example, graduate students may embed themselves in youth development programs, helping the educators with program activities while making observations of how youth develop assets over an extended period. Embedded assessment activities, such as youth journaling and conference presentations, also enable evaluators to gain insight into the feedbacks between participants and program context (Delia and Krasny 2018; Briggs et al. 2019).

Additionally, educators can assess their program to determine how well it is providing a context for youth development. For example, you might ask whether your program provides a safe space for youth of multiple ages, genders, socioeconomic classes, and ethnicities to express their ideas and feelings (Hinson et al. 2016).

15

HEALTH AND WELL-BEING

Highlights

- Health and well-being encompasses physical, cognitive, emotional, and social elements.
- Hedonic well-being is about pleasure and happiness, whereas eudaimonic well-being refers to reaching one's potential and finding meaning in life.
- Hope and coping strategies are important aspects of well-being in light of threats posed by climate change.
- Environmental education can partner with organizations focused on well-being, including those that serve people impacted by poverty and discrimination and displaced by conflict and climate change.
- Environmental education can foster hope and meaning in life through nature-based, outdoor adventure, civic ecology, and environmental action programs.

Spending time in nature enhances health and well-being—we know this from hundreds if not thousands of studies. Nature-based activities impact health and well-being through enabling physical exercise, social interactions, recovery from stress, and opportunities to freely explore one's surroundings. Nature-based physical and psychological challenges, hands-on stewardship, and being able to influence environmental policies also help us to find purpose or meaning in life, or eudaimonic well-being (Health Council of the Netherlands 2004;

Cervinka et al. 2011). Further, environmental education can instill hope and enable coping behaviors, which are critical in light of threats posed by climate change (Ojala 2012a, 2013, 2015; see table 15.1 for definitions of terms related to well-being). For a comprehensive review of health/well-being outcomes of activities in nature, we suggest the article "The Benefits of Nature Contact for Children" (Chawla 2015), Richard Louv's (2008) popular book *Last Child in the Woods: Saving Our Children from Nature-Deficit Disorder*, and the research reviews hosted on the Children & Nature Network and North American Association for Environmental Education websites (C&NN 2017; NAAEE 2018b).

What Are Health and Well-Being?

> **Much of the human search for a coherent and fulfilling existence is intimately dependent upon our relationship to nature.**
>
> (Kellert and Wilson 1993, 43)

The World Health Organization's definition of health emphasizes not just the absence of disease, but also physical, mental, and social well-being, and provides a starting point for thinking about environmental education outcomes (Chawla 2015; WHO 2018). Outdoor environmental education programs provide opportunities for increasing physical fitness, whereas programs taking place in school and community gardens can address healthy eating habits (Dyg and Wistoft 2018). Cognitive well-being includes memory, critical thinking, and judgment, whereas psychological well-being encompasses emotions, life satisfaction, self-esteem, and optimism (Lucas et al. 1996; Wells 2000, 2014; Wells and Rollings 2012). Social well-being can include social connections, social capital, and sense of community (McMillan and Chavis 1986; Sullivan et al. 2004; Krasny, Kalbacker, et al. 2013; Chawla 2015).

Psychological well-being is particularly important in environmental education. It can be considered from the hedonic perspective of satisfying desires and maximizing pleasure or happiness. It also can be considered from a eudaimonic perspective, which focuses on meaning in life resulting from following deeply held values and realizing one's fullest potential (Halama and Dedova 2007; Capaldi et al. 2014; McLellan and Steward 2015). The two perspectives are not necessarily at odds—for example, being outdoors brings great pleasure as well as meaning in life (Wolsko and Lindberg 2013; Chawla 2015).

The "capabilities approach" links eudaimonic well-being with health and social justice (Nussbaum 2011) and has been used to assess the benefits for

TABLE 15.1 Well-being definitions

WELL-BEING COMPONENT	DEFINITION
Well-being	
Eudaimonic	Well-being derived from finding meaning in life as a result of following deeply held values and realizing one's fullest potential (McLellan and Steward 2015)
Hedonic	Well-being derived from satisfying desires and maximizing pleasure or happiness (McLellan and Steward 2015)
Hope	
Constructive	Encompasses positive reappraisal (reframing worries about a problem to activate hope), faith in institutions and technology to address the problem, and faith in one's ability to improve problems (Ojala 2012a, 2015)
Unrealistic	Based on the denial of the severity of the problem (Ojala 2012a, 2015)
Coping	
Problem-focused	Entails searching for information about how one can solve a problem
Emotion-focused	Involves avoiding negative feelings, including by denying or not caring about a problem
Meaning-focused	Evokes positive emotions that aid in facing a difficult situation, while confronting rather than avoiding negative emotions

children of spending time in nature (Chawla 2015). In addition to addressing more commonly cited aspects of well-being such as physical health and forming attachments with other people, the capabilities approach includes the ability to connect with and express concern for animals, plants, and other aspects of nature (Nussbaum 2011). According to philosopher Martha Nussbaum, "To promote capabilities is to promote areas of freedom, and this is not the same as making people function in a certain way" (25). This suggests a potential connection to environmental action programs that seek to build youths' decision-making abilities (Stapp et al. 1996; Jensen and Schnack 1997; Volk and Cheak 2003; Wals et al. 2008; Schusler and Krasny 2010; Schusler 2014). The capabilities approach also bears similarities to social justice and other positive youth development approaches (Ginwright and Cammarota 2002; Catalano et al. 2004; see also chapter 14).

Nussbaum's (2011) first and last capabilities—being able to live a normal life and participate in the political process—are particularly salient when considering climate change. To further address climate change, environmental educators have begun to incorporate hope and coping strategies as components of well-being (Fritze et al. 2008; Ojala 2015). Hope encompasses goals, pathways for achieving those goals, and agency, or the capacity and motivation to use those pathways to reach desired goals (Snyder et al. 2018). Whereas people with low levels of hope "tend to catastrophize about the future, those with high levels of hope are able to think effectively about the future, with the knowledge that they, at times, will need to face major life stressors" (Snyder et al. 2018, 16).

Similar to attitudes (chapter 7) and sense of place (chapter 9), hope includes a cognitive aspect—beliefs about the future—as well as an affective component—positive feelings. Positive, and negative, emotions flow from our perception of how well we are achieving our goals (Ojala 2012a; Snyder et al. 2018). Overcoming impediments to reaching our goals yields positive emotions, and in this way, hope is associated with well-being (Snyder et al. 2018). Similar to self-efficacy (Bandura 1977; see also chapter 10), hope is goal directed. However, hope focuses on the general belief that one will initiate action, whereas self-efficacy refers to the belief in one's capacity to act in specific situations (Snyder et al. 2018).

Why Are Health and Well-Being Important?

> **At a time when the predictions from our most credible scientists are becoming increasingly grave, those involved in mental health promotion need to pay close attention to the relation between evidence, hope and action.**
>
> (Fritze et al. 2008, 9)

Parents, city governments, federal agencies, and universities have expressed intense interest in the role of nature in children's and adults' health. Some have gone so far as to start Nature Rx programs where people are prescribed time in nature to improve physical and emotional health (Rakow 2018). Environmental education can link with these efforts and serve as a reminder of nature's role in giving pleasure and meaning to life, or hedonic and eudaimonic well-being.

- A focus on health outcomes links environmental education to multiple societal concerns, ranging from social isolation to obesity to attention deficit disorder (Chawla 2015; C&NN 2017).
- Including health and well-being outcomes provides an opportunity for environmental education to partner with other organizations, including those serving people impacted by poverty and discrimination and displaced by conflict and climate change.
- Emotional health encompasses hope, which is increasingly relevant to environmental education in light of grave predictions about climate change (Fritze et al. 2008; Ojala 2015).
- Health and well-being outcomes can lead to environmental behaviors and actions.

How Do Health and Well-Being Lead to Environmental Behaviors?

> **It is only through the cultivation of this larger self that we are able to simultaneously experience great meaning and joy, as well as maintain a respectful and cooperative relationship with the natural world. In essence, the goals of conservationists and mental health professionals may become co-realized with this shift in identity and awareness.**
> (Wolsko and Lindberg 2013, 81)

Mainstream thought has often pitted environmental behaviors against individual well-being, claiming that reducing consumption leads to lower quality of life. In fact, well-being can be an outcome of environmental behaviors and collective action. Well-being also fosters environmental behaviors and actions through two pathways: connectedness to nature leading to well-being and environmental behaviors, and hope enabling productive coping strategies and environmental behaviors (figure 15.1).

Connectedness to nature is associated with both hedonic (pleasure seeking) and eudaimonic (meaning seeking) psychological well-being. Connectedness to

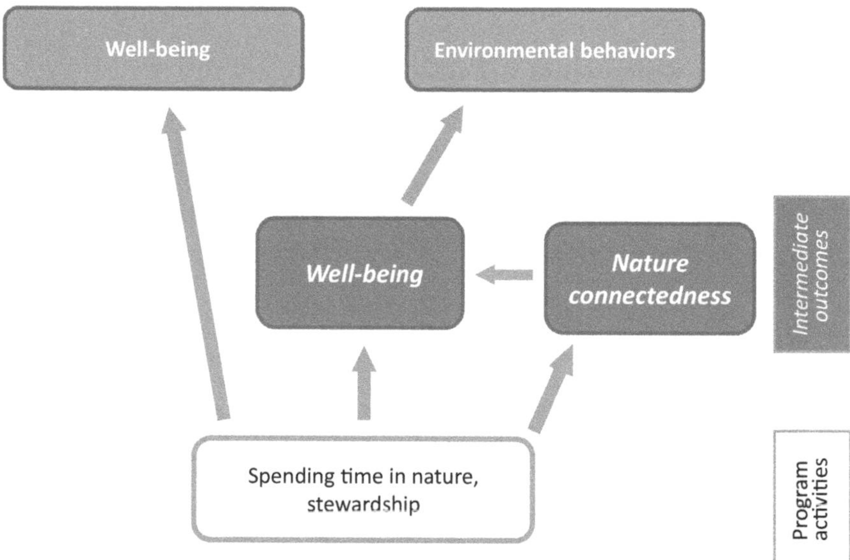

FIGURE 15.1 Well-being, including happiness, hope, and coping skills, enables youth to engage in environmental behaviors. Nature connectedness can lead to feelings of well-being, which is a desired outcome of environmental education.

nature is also associated with environmental behaviors (see chapter 8). In short, the same nature-based activities that enable connecting with nature also promote psychological well-being, including happiness, vitality, and life satisfaction (Wolsko and Lindberg 2013; Capaldi et al. 2014). For example, we find pleasure and meaning in nature walks and in planting a community garden, both of which are means to connect with nature.

Climate change can cause negative feelings including worry, sadness, anger, helplessness, and pessimism concerning the future (Stevenson and Peterson 2016). Constructive hope and productive coping strategies help people address the threat of climate change and associated negative emotions (Ojala 2013, 2015). Constructive hope encompasses three components and has implications for climate change education. First, positive reappraisal entails reframing worries about environmental problems into a means to activate hope. An example of positive reappraisal would be noting that despite the terrible consequences of climate change, there appears to be a growing awareness and willingness to act to reduce greenhouse gases. Second is faith in environmental organizations and technology to help address the problem. Third is faith in one's own ability to influence environmental problems in a positive direction. Constructive hope offers an alternative to an unrealistic hope based on the denial of the severity of the problem (Ojala 2012a, 2015).

Coping is a means to respond to threats to the well-being of oneself, of others, of future generations, and of nature (Ojala 2013; Snyder et al. 2018). Three types of coping strategies are used to respond to climate change and other environmental problems.

Problem-focused coping entails searching for information about how one can solve a problem. When individual or societal actions make a difference to the problem at hand, problem-solving coping is associated with hope and well-being, as well as with environmental behaviors. However, in children and adolescents, problem-solving coping also was associated with sad and anxious feelings, perhaps because those children who want to solve environmental problems also worried about climate change (Ojala 2012b, 2013).

Emotion-focused coping involves avoiding negative feelings, including by denying or not caring about a problem. Such denial and egocentric thinking, while helping to reduce anxiety when an individual cannot control a situation, is also used in situations where exerting control is possible. For example, conservatives may deny climate change in order to reduce threats to their cultural identity, even though it is possible to take actions to adapt to and mitigate climate change. Although such denial strategies may reduce anxiety, emotion-focused coping is not associated with longer-term well-being and does not lead to environmental behaviors or collective action among children, adolescents, and young adults (Ojala 2012a, b, 2013, 2015).

A third strategy, *meaning-focused coping*, evokes positive emotions such as hope, while confronting rather than avoiding negative emotions. It entails an individual acknowledging the problem, finding meaning and even benefits in a difficult situation, revising goals, and where appropriate turning to spiritual beliefs. For example, a person could acknowledge that climate change is a threat but also recognize hopeful aspects, such as the fact that more and more cities are taking action to reduce greenhouse gases. This type of coping is particularly helpful in evoking positive emotions while confronting a problem that cannot immediately or perhaps can never be resolved, like a terminal illness (Ojala 2013); it is associated with positive emotions, optimism, life satisfaction, and environmental behaviors in children (Ojala 2012b) and adolescents (Ojala 2013). Meaning-focused coping and constructive hope are particularly important in dealing with climate change because it is a long-term problem that cannot be solved by an individual acting alone, and where a means to recognize but buffer negative emotions is needed (Ojala 2012b, 2015).

Author Reflections

I am guilty of using denial as a coping strategy. I often avoid reading the news about climate change, unless it is good news, like a new renewable energy policy or some other innovation. I also search for hope, but find it in unusual ways—by picking up litter, even down to cigarette butts, on sidewalks and grass. To me each piece of trash or cigarette butt I pick up on my walk to work is a symbol of hope—hope that others won't litter, that they will treat our shared environment with respect. Thus my actions symbolically give me hope even as I recognize their seeming futility. But perhaps I am not alone—each little action sparks hope, and perhaps can be scaled up to larger collective actions (Krasny 2018).

How Can Environmental Education Foster Health and Well-Being?

> **The memory of planting a tree, of taking a pro-sustainability action in the past, could provide a source of well-being when recalled in later years.**
>
> (Waite et al. 2016, 57)

The research is clear—spending time in nature is good for our physical, cognitive, and psychological health and well-being. It can provide pleasure, or hedonic well-being, and meaning in life, or eudaimonic well-being. Further, spending time in nature with caring and respectful others can enhance social well-being. Thus, when environmental education provides opportunities for audiences to spend time in nature, it contributes to well-being on multiple fronts. But how else can environmental education contribute to well-being beyond simply providing time in nature? In this section, we focus on environmental education strategies to engender meaning in life or eudaimonic well-being, and to build hope and productive coping.

Meaning in Life

The simplest environmental education strategy for promoting meaning in life is allowing participants to spend time in nature (Howell et al. 2013), with the added component of allocating time for self-reflection. Outdoor recreation programs, such as hiking and canoeing, allow participants to find pleasure in and connect with nature, thus contributing to both hedonic and eudaimonic well-being (Wolsko and Lindberg 2013). Adventure education programs focus on physical challenges (e.g., backpacking, rock climbing, whitewater kayaking), which can lead to a sense of accomplishment and self-esteem. These programs also often incorporate "solo" experiences where participants spend time alone in nature to encourage reflection. As part of these and other wilderness or outdoor experiences, participants gain a perspective on how meaningfulness in life, as well as joy and happiness, are not always linked to consumption and everyday material comforts, but rather to the freedom to move about, to reflect, and to experience a sense of connection with nature. Whereas team building to conquer physical challenges is part of many outdoor programs, a focus on the intrinsic values of outdoor living is also essential in order for outdoor experiences to convey meaning in life (Sandell and Öhman 2010; D'Amato and Krasny 2011). In short, outdoor experiences can contribute to well-being through opportunities to have fun, conquer significant physical and mental challenges (often in cooperation with others), reflect, and experience the intrinsic value of nature.

Gardening, tree planting, and other stewardship opportunities similarly provide opportunities for finding meaning in life through connecting with nature and with others, as well as physical well-being outcomes as a result of eating healthy foods and getting exercise. Further, participants in school and community gardening and environmental stewardship programs find meaning through contributing to the environment and to their community (Krasny and Tidball 2015; Waite et al. 2016; Dyg and Wistoft 2018). Environmental stewards form memories that can be drawn on later in life and continue to foster psychological well-being (Waite et al. 2016).

Program participants also can contribute to their community and the environment, and thus find meaning, through programs where they engage in action to influence policy (Jensen and Schnack 1997; Volk and Cheak 2003; Schusler and Krasny 2010; Schusler 2014). By enabling youth to create a vision, discuss pathways to achieve their vision, and take action, environmental action programs can foster a sense of hope (Ojala 2015).

Hope and Coping

> **Programs should be able to significantly increase hopefulness if they foster sense of efficacy through providing imagery of what others are doing at both personal and community level.**
>
> (Li and Monroe 2017, 13)

To engender constructive hope, environmental educators should couple demonstrating respect for participants' negative emotions regarding climate change and other forms of environmental decline with a positive, solution-oriented communication style (Ojala 2015). As environmental educators are increasingly faced with students' and their own emotional reactions to environmental degradation (Fraser and Brandt 2013), their challenge is to evoke hope for the future through how they talk and act concerning climate change and other environmental issues. In short, educators act as role models for what students can achieve, in particular through invoking problem- and meaning-focused copying strategies (Ojala 2012b, 2013, 2015). While honoring and helping students process negative emotions, educators also can use negative emotions as teachable moments. Ignoring student emotions can lead to hope based on denial and thus unproductive coping strategies (Ojala 2015; Stevenson and Peterson 2016).

Once having recognized participants' emotions, programs can engender hope through building self- and other forms of efficacy (see chapter 10), including through allowing students mastery experiences and learning about what others are doing as individuals and as groups (Li and Monroe 2017a).

Assessing Health and Well-Being

For cognitive and affective well-being, researchers have developed and tested numerous surveys. Here we focus on three scales developed for children—a survey to test cognitive, affective, and social well-being; a survey to measure climate change hope; and a survey to measure coping in relation to climate change (see appendix). Apps that help you keep track of physical activity, food consumption, and even cortisol in your saliva as an indicator of stress can also be used as embedded assessments.

In a survey developed to measure well-being, children are asked to indicate their level of agreement with statements in four categories: interpersonal well-being (e.g., "I feel cared for"); life satisfaction (e.g., "I feel there is lots to look forward to"); perceived competence (e.g., "I feel I can deal with problems"); and negative emotion (e.g., "I feel worried") (McLellan and Steward 2015).

A climate change hope scale for use with children encompasses two factors: willpower, or the belief one is able to reach a goal or overcome a problem, and "waypower," or the ability to generate pathways to overcome a problem. Because climate change requires that individuals and society take action, this scale includes both individual and collective willpower and waypower. Likert scale questions for individual willpower and waypower include "I know that there are a number of things that I can do to contribute to global warming solutions" and "I am hopeful about global warming because I can think of many ways to resolve this problem." For collective willpower and waypower, students are asked to indicate their level of agreement with statements such as, "I believe people will be able to fix global warming" and "Because people can change their behavior, we can influence global warming in a positive direction" (Li and Monroe 2017a).

Environmental educators may also want to assess participants' climate change coping strategies, which vary in their relationship to well-being and environmental behaviors. In a survey designed for children, meaning-focused coping, which is associated with environmental behaviors and well-being, was assessed by asking participants their level of agreement with statements such as "I have faith in people engaged in environmental organizations to address climate." For problem-focused coping, which is associated with environmental behaviors but also with negative emotions such as worry, statements include "I talk with my family and friends about what one can do to help." Finally, emotion- or denial-based coping, which may reduce anxiety among children but is generally not associated with long-term well-being or environmental behaviors, is measured with statements that include "I think the problem is exaggerated" (Ojala 2012b).

Conclusion

RESILIENCE: ADAPTATION AND TRANSFORMATION

Highlights

- Resilience captures how humans, communities, ecosystems, and social-ecological systems bounce back, adapt, and transform in the face of ongoing change and major hardships or catastrophes.
- Resilience recognizes the need to incorporate small perturbations and large disasters, including those caused by climate change, in environmental management, planning, policy, and education.
- Biological diversity and ecosystem services; systems thinking and the ability to anticipate and learn from unexpected dynamics; the ability to manage slow processes like climate change; cultural diversity, civic participation, social capital, and polycentric governance—all these factors contribute to social-ecological systems resilience.
- Environmental education can foster resilience through environmental action programs where youth and adults partner with nonprofits and government agencies to steward public spaces, thus providing ecosystem services and opportunities to develop social capital, systems thinking, and other youth and community assets.

In 2012, Hurricane Sandy struck New York City, causing massive flooding and destruction and sparking a fire that leveled more than one hundred homes. At the time of the superstorm, New York City housed hundreds of environmental organizations engaged in stewardship, advocacy, and education (Svendsen and

Campbell 2008; Kudryavtsev et al. 2012). Because the storm exemplified what appeared to be increasingly more powerful climate-related disasters, the question arose: How would the educational efforts of these environmental organizations change as a result of the storm (DuBois and Krasny 2016; Krasny and DuBois 2016)? It turns out that despite their diverse approaches to education, New York City environmental educators described their response to the storm using a common term: resilience.

In this concluding chapter, we focus on individual, community, and social-ecological systems resilience. By incorporating notions of how people, communities, and social-ecological systems respond to change, resilience enables environmental educators to integrate outcomes across different levels. In addition to environmental quality and sustainability (see chapter 3), social-ecological resilience can be considered as an ultimate outcome of environmental education that gives equal weight to environmental and societal concerns. Resilience challenges linear theories of change by taking into account complex system dynamics or feedbacks, lending support to the idea that intermediate outcomes can both lead to behaviors and be an outcome of behaviors and action. Importantly, by focusing on how people and systems change, resilience offers insights into how society can address rising sea levels, devastating fires, destructive storms, and other changes brought about by climate change. Finally, social-ecological resilience recognizes that multiple organizations and actors—including environmental educators—are essential to bring about the needed adaptations and transformations to address climate change and other gradual shifts and major disasters.

What Is Resilience?

> **Resilience thinking is about how periods of gradual changes interact with abrupt changes, and the capacity of people, communities, societies, cultures to adapt or even transform into new development pathways in the face of dynamic change.**
>
> (Folke 2016, 2)

What did the New York City environmental educators mean by resilience, given the term's multiple definitions? The educators used "resilience" to describe how they were helping program participants, communities, ecosystems, and social-ecological systems respond to the storm and prepare for future disaster (DuBois and Krasny 2016). In this way, their use of the term spanned how it is used by psychologists, sociologists, and environmental scientists (table 16.1).

TABLE 16.1 Resilience definitions

TYPE OF RESILIENCE	DEFINITION
Psychological	The processes of, capacity for, or patterns of positive adaptation during or following exposure to adverse experiences that have the potential to disrupt or destroy the successful functioning or development of the person (Masten and Obradovic 2008)
Community	The ability of communities to cope with and recover from external stressors resulting from social, political, and environmental change (CARRI 2013)
Ecological	The magnitude of disturbance that a system can experience before it moves into a different state with different controls on structure and function (Holling 1973)
Social-ecological systems	The capacity of a social-ecological system to continually change, adapt, or transform so as to maintain ongoing processes in response to gradual and small-scale change, or transform in the face of devastating change (Berkes et al. 2003)

For psychologists, resilience refers to how people who face extreme hardships are able to "bounce back" and go on to live productive lives, in part by drawing on positive emotions rather than focusing on problems (Luthar et al. 2000; Bonanno 2004). After disasters like Hurricane Sandy, community gardening and other greening activities can spur positive emotions, as well as enhance cognitive capacity and community engagement, thus reducing distress and fostering psychological resilience (Okvat and Zautra 2014). As one New York City environmental educator described her definition of resilience,

> But what is that element to allow them to withstand major trauma and how do we help to prepare and equip young people to have those skillsets? . . . It is sort of the grit, and the idea of the resilience of the human capacity. (New York City environmental educator, December 5, 2014, quoted in DuBois and Krasny 2016)

Another environmental educator was largely concerned about community resilience:

> Having green space and an urban farm where people can actively do work is in itself a form of resiliency. Not in a direct environmental way, but in a community way. So if people in the city have a space where they can come to do work in greenspace—and have a connection to it and really learn to care about it—then that is for the resiliency of the population. And over time, the more people fostering

> the care for these spaces the more these spaces will exist. (New York City environmental educator, December 16, 2014, quoted in DuBois and Krasny 2016)

A third educator thought about resilience in terms of a shift from a focus on biodiversity to emphasizing ecosystem services, a notion consistent with ecological resilience.

> Instead of increasing diversity—[we are] looking at the structure and diversity of forest to withstand intense storm events and recover without degrading to intensive vineland. [It] provides ecosystem services in terms of storm water management and carbon sequestration . . . ecosystem services that are more critical with increased precipitation. Intense environmental stresses that we predict and are already seeing. A natural restoration project—but have added a layer around the increased storm events and resilience of forest. (New York City environmental educator, December 9, 2014, quoted in DuBois and Krasny 2016)

Finally, educators described how they fostered community and ecological resilience, or social-ecological resilience. This commonly occurred through participation in civic ecology practices such as community gardening or volunteer tree planting (Krasny and Tidball 2015).

> So I think in all of our work we don't try to make it just about the people or just about the trees, kind of bridging the two aspects through community engagement. (New York City environmental educator, December 4, 2014, quoted in DuBois and Krasny 2016)

Regardless of whether we are referring to individuals, communities, ecosystems, or social-ecological systems, resilience broadly refers to the ability to adapt to ongoing change and to transform after major disruptions. Resilience leverages a community's capacity to respond to disturbances and thus offers a counterpoint to a focus on a community's vulnerability (Norris et al. 2008). Social-ecological resilience emphasizes how social and ecological processes are tightly linked. Important for environmental education in an age of climate change and other major disruptions, social-ecological resilience draws attention to a system's ability to adapt to small ongoing changes, such as increases in nutrients running off into streams, as well as to transform following major catastrophes like Hurricane Sandy or California wildfires that level residential and forested areas alike.

Whereas "resilience" is often used in the positive sense, the term has been criticized as implying that individuals and social-ecological systems will simply adapt to deteriorating conditions like poverty or climate change. Such ongoing adaptation can counter efforts to address larger structural issues, such as economic inequality or fossil-fuel-dependent energy policy (Nadasdy 2007; Lotz-Sisitka et al. 2015). While the notion of resilience as simply "bouncing back" is indeed problematic, many definitions of resilience focus not only on adaptation and bouncing back, but also on the capacity of systems to transform after a major disturbance or catastrophe. This is because many of the same capacities that enable a system to adapt also are needed for transformation. Further, by emphasizing the notion that systems are always changing, often in unexpected ways, resilience can help us anticipate and manage for change. In short, a resilient system not only responds to change but also takes advantages of opportunities brought about by disruptions to create new, transformative systems that benefit humans and other organisms (Gunderson and Holling 2002; Folke et al. 2003; Biggs et al. 2015; Folke 2016). An example comes from a city in New York State, where for years riverside neighborhoods tried to rebuild or adapt after repeated flooding. Eventually, the US Federal Emergency Management Agency and citizen groups began working together to resettle residents and transform the shoreline social-ecological system to green space that will provide recreational opportunities, carbon sequestration, and buffering against floodwaters (Binghamton 2016).

Resilience offers an alternative to sustainability, another ultimate outcome of education that integrates social and environmental concerns. Much has been written about environmental education and sustainability, and UNESCO supports "Education for Sustainable Development" as an alternative to environmental education that incorporates issues of social and economic justice (UNESCO 2002, 2007, and n.d.; Monroe 2012; Wals 2012; see also introduction). I prefer to focus on resilience for two reasons. First, addressing change through adaptation and transformation is critically important and is core to social-ecological resilience scholarship and definition. Resilience scholars have gone so far as to invent a new term, "panarchy," which explores how cycles of adaptation and transformation at local, regional, and global scales both constrain and spur transformation in each other (Gunderson and Holling 2002). Second, resilience is a term that often resonates with educators and the broader public, perhaps because we inherently understand our own struggles and psychological resilience, and can intuitively relate these experiences to the resilience of communities, ecosystems, and social-ecological systems (Krasny et al. 2010, 2011; Lundholm and Plummer 2010; Sterling 2010).

Why Is Resilience Important?

> **The key question at this point is what kinds of education and learning experience are appropriate for a world where surprise and unaccustomed levels of change will likely become major features of our lives.**
>
> (Sterling 2010, 521)

The notion of resilience forces us to recognize the ubiquity of change, including ongoing small changes and major hardships and disruptions. These changes impact environmental education participants, the communities where they live, and the social-ecological systems that we set out to enhance.

- Resilience thinking is desperately needed as we face potentially threshold-level (tipping point) changes related to climate change, plastics pollution, and other disruptions.
- Resilience suggests ways in which we can alter "maximize yield" or reductionist thinking, with the goal to foster longer-term sustainability. For years, our attempts to manage nature for maximum benefit for humans have led to unanticipated consequences. An example comes from how applying nitrogen, phosphorus, and other nutrients to increase crop production has resulted in runoff of chemicals and unexpected harmful algal blooms in lakes.
- Resilience thinking suggests that humans are part of systems, not simply managers separate from nature. The term "social-ecological resilience" reinforces the idea that humans are intimately connected to ecosystems and that we manage integrated rather than separate human and natural systems.
- In situations where there are major disruptions, for example massive flooding in cities, resilience thinking can help us design novel and sometimes transformative solutions. An example comes from New York City's East Side Coastal Resiliency Project, which intends to use berms and other types of innovative green infrastructure "to integrate flood protection into the community fabric, improving access to the waterfront rather than walling off the neighborhood" (New York City 2018).
- Individuals in our programs may face hardships at home and in school or community, and may feel vulnerable upon hearing about or experiencing climate change–related disasters. Environmental education can borrow approaches from psychological and community resilience to help youth.

- Educators can apply notions of resilience, change, and vulnerability to their professional and personal lives. How are we as environmental educators adapting or even transforming our programs given climate change and associated hardships and disasters? And in light of the depressing news about climate change, how can we address our participants' and our own well-being, for example through forming networks to support each other and through taking action (Fraser and Brandt 2013)?

What Factors Foster Social-Ecological Resilience?

Resilient social-ecological systems generally have a set of attributes like biodiversity and ecosystem services, systems thinking, social capital, and civic participation in stewardship and governance (Walker and Salt 2006; Biggs et al. 2012; Biggs et al. 2015; Gunderson et al. 2018; see figure 16.1). Environmental education can play a role in fostering resilient system attributes; for example, community or school gardening programs provide ecosystem services and can enhance systems thinking and social capital (Tidball and Krasny 2009; Krasny, Lundholm, et al. 2013). Below we provide an overview of factors that foster social-ecological resilience—that is, factors that enable a system to adapt and transform.

Biological Diversity and Ecosystem Services

Resilient social-ecological systems have high levels of biodiversity. Community or school gardening, invasive species removal, tree planting, and other

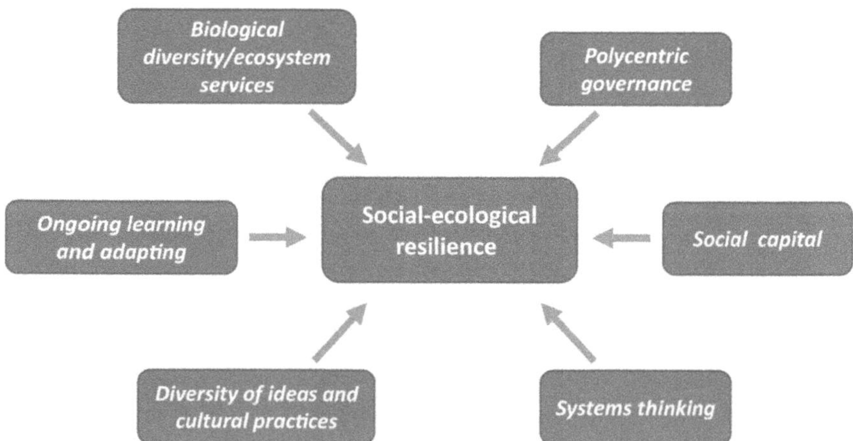

FIGURE 16.1. Multiple factors, many of them outcomes of environmental education, contribute to social-ecological resilience.

environmental education programs situated in civic ecology practices can contribute to such diversity (Krasny, Lundholm, et al. 2013). Biodiversity in turn provides ecosystem services, such as food, pollination, and erosion control, as well as opportunities for connecting to nature. When multiple species and varieties are providing the same ecosystem service, the system is better able to buffer disturbances. For example, having multiple varieties of apples may enable an orchard to withstand an insect infestation or late spring frost, which kills some varieties but leaves others intact. And plants along a coastline whose roots penetrate to different depths may offer more resistance to flooding than a uniform row of a single species (Walker and Salt 2006).

Managing Slow Processes

Humans are good at responding to "fast variables" like changes in the weather. However, we are not as good at responding to "slow variables," or slow-moving processes like climate change and population growth. Managing for resilience entails taking into account these slow processes (Walker and Salt 2006).

Managing Feedbacks

We often think of environmental education programs as leading to an outcome; for example, a community gardening education program will increase trust among youth and elders. But by building trust we are also changing our community of gardeners, which opens up opportunities for further change. A group of youth and adults who trust each other are more likely to continue working in their garden, or even to advocate for the protection of urban green space. This process is called a desirable feedback. An undesirable feedback might be an invasive species that reduces habitat for native species, thus allowing the invader to spread. Managing for resilience means anticipating feedbacks, or instances where a change in a particular variable or process reinforces subsequent changes of the same variable or process (Biggs et al. 2012; Tidball et al. 2017). Feedbacks are prevalent in environmental education—for example, engaging in an environmental behavior may lead to further engagement in that behavior.

Complex Adaptive Systems Thinking

Environmental education has traditionally promoted systems thinking, which involves shifts in focus from understanding the parts of systems, such as particular objects or content, to understanding whole systems, including their components, dynamics, and functions (Capra 2007; Hmelo-Silver et al. 2017; see also chapter 6). A resilience perspective is consistent with systems thinking owing to its focus on system complexity, adaptation, and transformation (Biggs et al. 2012). In contrast, reductionist or linear thinking focuses narrowly on one aspect of the system. For example, linear thinking would focus on how adding fertilizers increases corn production—which, while true, fails to account for other processes and less predictable outcomes, such as excess nutrients running off into lakes. For a time the lake ecosystem absorbs the excess nutrients, so we don't notice any change in the lake. But suddenly we find the lake covered by harmful blue-green algae. In this situation, the lake ecosystem has crossed an unanticipated threshold, or tipping point, and is no longer able to function as a habitat for desired fish species and as a setting for outdoor recreation. When managers acknowledge the uncertainty about the ways in which social-ecological systems behave, they can accept the need for ongoing learning and experimentation. Adaptive learning or learning-through-experience in management and in stewardship practices is important (Biggs et al. 2012).

Cultural and Knowledge Diversity

When people from different backgrounds participate in environmental stewardship and formulating policy, they contribute different perspectives and knowledge. These perspectives and knowledge can be applied to devising novel ways of adapting to change and transforming dysfunctional systems (Walker and Salt 2006).

Participation, Learning, and Social Capital

> **The paradox of education is that it is seen as a preparation for the future, but it grows out of the past. In stable conditions, this socialization and replication function of education is sufficient: in volatile conditions where . . . the future will not be anything like a linear extension of the past, it sets boundaries and barriers to innovation, creativity, and experimentation.**
>
> (Sterling 2009, 19, quoted in Sterling 2010)

Youth and adult participation in decision making and hands-on stewardship can help build connections and trust (that is, social capital; see chapter 13) and can provide diverse perspectives on stewardship outcomes and how to achieve them. Often participation encompasses citizen science or other forms of monitoring biodiversity and ecosystem services, and thus can contribute much-needed information about management practices and when they need to be adjusted.

From an environmental education point of view, participation in stewardship and monitoring can enhance understanding of social-ecological system dynamics (see chapter 6) and build youth assets (chapter 14). When diverse stakeholders including scientists and on-the-ground resource managers participate in stewardship decisions, actions, and outcomes monitoring, what they learn can be readily translated to ongoing improvements in stewardship practices (Walker and Salt 2006).

Polycentric Governance Systems

Whereas government refers to policies set by formal government agencies, *governance* is about how multiple actors, including NGOs, the private sector, and government agencies, deliberate and make decisions about policies. Governance systems with NGO, private industry, and government actors are referred to as "polycentric" (many centers). Similar to how groups of people holding diverse perspectives think of novel ways to adapt and even transform social-ecological systems, governance actors from different sectors offer multiple options for adaptation and transformation and thus contribute to the resilience of social-ecological systems (Ostrom 2010b; Gunderson et al. 2018). In the Chesapeake Bay watershed, local governments and citizens' groups monitor individual tributaries and help make decisions about these smaller bodies of water, while large nonprofits such as the Chesapeake Bay Foundation work with federal and state governments to formulate policies for the entire watershed. Matching such decision making to the scale of the resource (e.g., small tributary, large watershed) and involving multiple government and civil society actors can foster resilience by allowing for meaningful participation and learning and for one group to take up the slack when another group changes direction. For example, when the national government moves away from clean water and climate change policies, regional or local government and citizens groups emerge and can work to ensure sound environmental policies and practices.

How Can Environmental Education Foster Resilience?

> **Education not only may be *about* resilience as a concept, but also may guide students in *fostering* resilience within the watershed, neighborhood, or other social-ecological system in which they work and live.**
>
> (Krasny et al. 2009, 3, italics in original)

Environmental education can teach about social-ecological systems and how to enhance their resilience. It also can foster attributes of resilient systems such as biodiversity, ecosystem services, and social capital. Finally, environmental education organizations can become part of larger environmental governance systems, which play a role in the ability of communities to devise innovative approaches to address ongoing and catastrophic change.

While recognizing that knowledge does not necessarily lead to change, understanding resilience principles can be one step in the training of conservation scientists and stewards or simply in creating critical thinkers. Courses intended to help students understand resilience often use experiential case studies and project-based learning, where students apply what they learn about integrated social-ecological systems to a real-world problem or case (Krasny et al. 2009). In the massive open online course (MOOC) on civic ecology, students learned about resilience and related ideas and then applied these principles to a local civic ecology practice (Krasny and Snyder 2016). Educators, especially those working with students coming from different backgrounds, offer guidelines for respectful discussions where students can share their experiences and perspectives and develop social connections. Incorporating a reflective writing component in these hands-on programs can deepen understanding.

Just as resilience thinking links human and natural systems, environmental education can link individual resilience and social-ecological resilience (McPhearson and Tidball 2013). Environmental educators are aware that unexpected events and challenges can lead to transformational learning (Mezirow 2000; D'Amato and Krasny 2011; O'Sullivan 2002), and social-ecological resilience suggests that major disturbances can spur learning and novel approaches to environmental management (Gunderson and Holling 2002). Further, learning by doing—that is, learning as part of trying out different management approaches—is critical to the ability to adapt and transform management practices based on the outcomes of various management schemes (Berkes 2004; Armitage et al. 2008). When youth in environmental education programs have opportunities to learn through authentic experiences, like stewarding plants and soils in a community garden, they may also develop self-efficacy and other assets integral to psychological resilience (Sterling 2010).

By fostering an understanding of resilience principles and of diverse perspectives and by creating social capital, case study and project-based learning incorporate aspects of learning *about* resilience and *for* resilience (Sriskandarajah et al. 2010). For example, learning that takes place through environmental action programs that are embedded in civic ecology practices can enhance learning about complex systems, while also contributing to the civic ecology practice itself. These efforts in turn contribute to the social-ecological resilience of a neighborhood by catalyzing social capital, enhancing ecosystem services, and incorporating diverse perspectives in decision making about the civic ecology practice and related physical resource (e.g., bioswale garden) (Walker and Salt 2006; Krasny et al. 2009; Krasny and Roth 2010; Krasny, Lundholm, et al. 2011; Monroe and Allred 2013; Krasny and Tidball 2015). Environmental action programs that include modeling activities, in which youth and other stakeholders diagram system components and dynamics (e.g., linear pathways and feedbacks among system components), can further facilitate learning about systems and how they work (Biggs et al. 2011; Smith et al. 2015; Gray et al. 2017). In short, environmental education can contribute to resilience by facilitating environmental action embedded in civic ecology and other collective stewardship practices, and by including opportunities for respectful sharing of knowledge and perspectives.

When environmental action programs become embedded in stewardship partnerships among several organizations, they become part of polycentric governance systems. Environmental education organizations, working in partnership with government, schools, NGOs, and the private sector, can play a role in determining and implementing local and even national environmental policies.

In short, designing environmental education to enhance social-ecological resilience involves situating learning in stewardship action to enhance biodiversity and ecosystem services, social capital, and polycentric governance. Such programs should involve a series of scaffolded learning activities for acquiring knowledge, building social support, and, where appropriate, making collective decisions over extended periods (Chawla and Cushing 2007; Biggs et al. 2015; Shihui et al. 2018). By paying heed to positive youth development (see chapter 14), such programs can also foster psychological resilience among participants who have experienced hardships, including those associated with climate change disasters.

Assessing Resilience

The Resilience Alliance has published a workbook that guides communities through a series of steps to assess local social-ecological resilience. These steps involve describing the social-ecological system, including its scale, key issues of concern, dynamics such as potential tipping points and ongoing interactions, and governance systems (Resilience Alliance 2010).

Determining a neighborhood's, city's, or other social-ecological system's resilience can be time-consuming (Resilience Alliance 2010), and thus environmental education organizations may want to partner with city government or community or environmental organizations whose main focus is enhancing local resilience. Environmental educators can also measure factors known to be present in resilient social-ecological systems (Walker and Salt 2006; Tidball and Krasny 2010; see also appendix). For example, protocols have been developed to measure ecosystem services (Krasny, Russ, et al. 2013), trust and social capital (Krasny, Kalbacker, et al. 2013; Ardoin, DiGiano, O'Connor, and Podkul 2016) and systems thinking outcomes of environmental stewardship and education programs (Smith et al. 2015; DuBois et al. 2017).

I hope you have enjoyed your journey exploring the myriad pathways by which environmental education can contribute to environmental quality, positive youth development, human health and well-being, and even hope and resilience in the face of climate change. Whether your pathway be through social capital or social norms, environmental identity or place identity, nature connectedness or critical thinking, self-efficacy or school efficacy, action-related or effectiveness knowledge, or simply choice architecture to make the green choice the easy choice, keep in mind that an active research community is discovering more effective ways to reach our common goals. And don't be shy about contacting researchers—they may be happy to discuss your questions and share their latest findings, or even be looking for partnerships with environmental educators to help ground their thinking in practice.

Finally, although much of this book has focused on how cognitive and affective factors lead to environmental behavior and collective action, we have also touched on how engaging in behaviors and actions can foster and reinforce norms, social capital, systems thinking, and other intermediate outcomes. Sometimes, you may want to simply engage your participants in the behavior or collective action you hope they will adopt for the long term. While participants engage in the behavior, you can seize opportunities to foster learning and reinforce positive emotions as they emerge. These intermediate outcomes—learning and emotions—may in turn lead to future behaviors. In short, keep in mind the feedbacks between doing and thinking and feeling—or behaviors/actions and cognitive/affective outcomes—as you develop your theories of change and plan your activities. Just as in life, we do not always find a linear pathway to reach our goals, yet we encounter myriad and sometimes unanticipated pathways to choose and to learn from.

Appendix

SURVEY INSTRUMENTS FOR ASSESSING ENVIRONMENTAL EDUCATION OUTCOMES

Survey Instruments Included in Appendix

Chapter 4. Environmental Behaviors

Pro-environmental Behavior Survey

Chapter 5. Collective Environmental Action

Environmental Action Scale

Chapter 6. Knowledge and Thinking

Environmental Knowledge Scales (System, Action-Related, and Effectiveness Knowledge)

Systems Thinking Scale

The Ultimate Cheatsheet for Critical Thinking

Chapter 7. Values, Beliefs, and Attitudes

Values Scale

New Ecological Paradigm

Environmental Attitude Inventory

Chapter 8. Nature Connectedness

Connectedness to Nature Scale

Connection to Nature Scale for Children

Nature Relatedness Scale

Chapter 9. Sense of Place
Sense of Place Scale

Chapter 10. Efficacy
Self-Efficacy and Collective Efficacy
Group, Participative, and Individual Efficacy
Political Efficacy (Internal and External)
School Efficacy

Chapter 11. Identity
Environmental Identity Scale

Chapter 12. Norms
Injunctive Social Norms
Descriptive Social Norms
Personal Norms

Chapter 13. Social Capital
Social Capital Survey for Youth

Chapter 14. Positive Youth Development
Active and Engaged Citizenship

Chapter 15. Health and Well-Being
Well-Being Scale
Climate Change Hope Scale
Climate Change Coping Scale

Chapter 4. Environmental Behaviors

Pro-environmental Behavior Survey (Larson et al. 2015, 118)

Note. This survey was developed for rural residents; you may want to adapt it for urban and suburban participants. Questions include individual behaviors and collective action.

Question. How often do you participate in these behaviors?

Scale. 1–5 (never—very often)

Conservation lifestyle

CL1. Recycled paper, plastic, and metal

CL2. Conserved water or energy in my home

CL3. Bought environmentally friendly and/or energy-efficient products

Land stewardship

LS1. Made my yard or my land more desirable for wildlife

LS2. Participated (provided data) in a wildlife study

LS3. Volunteered to improve wildlife habitat in my community

Social environmentalism

SE1. Talked to others in my community about environmental issues

SE2. Worked with others to address an environmental problem or issue

SE3. Participated as an active member in a local environmental group

Environmental citizenship

EC1. Voted to support a policy/regulation that affects the local environment

EC2. Signed a petition about an environmental issue

EC3. Donated money to support local environmental protection

EC4. Wrote a letter in response to an environmental issue

Chapter 5. Collective Environmental Action

Environmental Action Scale (Alisat and Riemer 2015, 19)

Note. You may need to adapt the questions depending on what kinds of collective action are allowable in your context (e.g., collective action may be possible within a school but not directed at government officials).

Question. In the last six months, how often, if at all, have you engaged in the following environmental activities and actions?

Scale. 0–4 (never—frequently)

1. Educated myself about environmental issues (e.g., through media, television, internet, blogs, etc.)
2. Participated in an educational event (e.g., workshop) related to the environment
3. Organized an educational event (e.g., workshop) related to environmental issues
4. Talked with others about environmental issues (e.g., spouse, partner, parent(s), children, or friends)
5. Used online tools (e.g., YouTube, Facebook, Wikipedia, MySpace Blogs) to raise awareness about environmental issues
6. Used traditional methods (e.g., letters to the editor, articles) to raise awareness about environmental issues
7. Personally wrote to or called a politician/government official about an environmental issue
8. Became involved with an environmental group or political party (e.g., volunteer, summer job, etc.)
9. Financially supported an environmental cause
10. Took part in a protest/rally about an environmental issue
11. Organized an environmental protest/rally
12. Organized a boycott against a company engaging in environmentally harmful practices
13. Organized a petition (including online petitions) for an environmental cause
14. Consciously made time to be able to work on environmental issues (e.g., working part time to allow time for environmental pursuits, working in an environmental job, or choosing environmental activities over other leisure activities)
15. Participated in a community event that focused on environmental awareness

16. Organized a community event that focused on environmental awareness
17. Participated in nature conservation efforts (e.g., planting trees, restoration of waterways)
18. Spent time working with a group/organization that deals with the connection of the environment to other societal issues such as justice or poverty

Chapter 6. Knowledge and Thinking

Environmental Knowledge Scales (Roczen et al. 2014)

Note. Includes system, action-related, and effectiveness questions. Scale available for download at http://journals.sagepub.com/doi/suppl/10.1177/0013916513492416/suppl_file/online_appendix_environmental_knowledge_items.pdf. Correct answers are indicated by bold text. Some questions are country-specific and will need to be adapted for your context.

System Knowledge

TRUE/FALSE QUESTIONS

SYS1: Oxygen is generated during forest fires. (**false**)

SYS2: Europe is the continent most affected by the hole in the ozone layer. (**false**)

SYS3: The sea level would rise by 80 m if all polar ice masses melted completely. (**true**)

SYS4: Young children who have frequent contact with animals are more susceptible to allergies later on. (**false**)

SYS5: When wind energy is converted, no CO_2 is emitted. (**true**)

SYS6: Ozone naturally occurs in forests to a larger extent than in nonforested areas. (**false**)

SYS7: If all ozone-destroying emissions were eliminated right now, it would take 100 years for almost complete regeneration of the ozone layer. (**true**)

SYS8: Solar energy is unlimitedly available. (**true**)

SYS9: The "El Niño" phenomenon is a direct consequence of global warming. (**false**)

SYS10: The vegetation of the hills and mountains of Bavaria is extremely resistant to external influences and even survived the last ice age. (**false**)

SYS11: When coal is converted into energy in a conventional power plant, a quarter of the energy is lost. (**false**)

SYS12: As a rule, clear lakes are not polluted with harmful substances. (**false**)

SYS13: If the concentration of atmospheric CO_2 was doubled, the global mean temperature would rise by about 5° Celsius (9° Fahrenheit). (**false**)

MULTIPLE-CHOICE QUESTIONS WITH ONE CORRECT ANSWER (IN BOLD)

SYS14: What does the abbreviation CO_2 stand for?

carbon dioxide

carbon monoxide

greenhouse effect

SYS15: Some devices, such as calculators, work with an environmentally friendly form of energy. What is it called?
solar energy
rechargeable batteries
wind energy

SYS16: Where are the tropical rain forests located?
South Africa
in a wide belt around the equator
Australia / New Zealand

SYS17: Which of these countries has the largest contiguous areas of forest?
The Netherlands
Spain
Brazil

SYS18: Forests bind . . . for a long time.
oxygen
carbon
ozone

SYS19: Which of the following kinds of energy is renewable?
solar energy (e.g., solar cells)
nuclear power
wind power

SYS20: Why is acid rain damaging to trees?
When the rain is deposited on the leaves, plants are not able to photosynthesize anymore.
Acid rain causes a displacement of minerals that are important for the plants.

SYS21: Where does most of the cellulose for German paper mills come from?
exclusively from German forests
from trees that have been planted for that purpose
from different indigenous forests around the world, e.g., from Canada, Russia, or Brazil

SYS22: On clear nights, why does it get colder toward the morning?
because a clear night sky supplies more cold than a cloudy sky
because the earth radiates heat, and there is no cloud cover to retain it
because the earth absorbs heat

SYS23: Why is CO_2 a problem?
CO_2 damages many species of plants.
CO_2 contributes to global warming.
CO_2 is poisonous to many microorganisms.
Levels of CO_2 are decreasing in the atmosphere.

SYS24: If trees are burned, . . . is produced.
CFC (chlorofluorocarbon)
oxygen
nitrogen
CO_2

SYS25: What does "sustainable forestry" mean?
The forest is used as effectively as possible.
Forestry that has not yet achieved the most up-to-date status.
Only as much wood as can be reforested is taken.

SYS26: In a humid climate (such as Bavaria), how long does it take for 10 cm (4 inches) of soil to form?
50 years
150 years
1,000 years

SYS27: Today's forestry is based on which principle?
the green guideline
persistency
sustainability

SYS28: Global warming also has an effect on the Gulf Stream that will affect Europe. What is this effect?
The Gulf Stream will possibly lead to additional warming of the climate.
The Gulf Stream will possibly collapse, which will lead to a strong cooling of the climate.

MULTIPLE-CHOICE QUESTIONS WITH MULTIPLE ANSWERS (IN BOLD)

SYS29: Solar energy can be used for . . .
heating water.
heating rooms.
cogeneration with fridges.
generating electricity.

SYS30: Why is paper bleached?

to get rid of the brown color of the wood

to save money

because people prefer white paper

SYS31: What are the protective functions of the forests? They protect against . . .

erosion.

radioactive contamination of the ground.

inundations.

SYS32: What are characteristics of fossil energy (such as coal and oil)?

They developed during the last 100 years.

During the conversion, CO_2 is released.

They are available only in limited quantities.

It took only 10 years for large-scale industrial exploitation to exhaust them.

SYS33: Where does groundwater come from?

It is very old and is no longer being formed.

It comes to the earth's surface from deep geological layers.

Seepage of rainfall into the ground.

Seepage through the beds of rivers and lakes.

SYS34: What causes wind?

the thrusting of the clouds

temperature differences

differences in air pressure

ocean currents

SYS35: What is unique about the tropical rain forest?

its biodiversity

its fertile soils

the absence of seasons

SYS36: What are problematic issues with ozone?

Ozone in the upper atmosphere is damaging because it reduces ultraviolet light from the sun.

Ozone damages the respiratory systems of people and animals.

Ozone reduces plant growth.

SYS37: During photosynthesis . . .

carbohydrates and oxygen are produced.

CO_2 is absorbed and oxygen is released.

light is converted into energy.

SYS38: What are the reasons for the destruction of the rain forest?
rich countries' demand for meat
industrial nations' demand for paper (e.g., Germany)
tourism
growing the base product for bio-diesel, for example, rapeseed

Action-Related Knowledge

TRUE/FALSE QUESTIONS

ACT1: All propellant gases in spray cans contribute to the greenhouse effect. **(false)**
ACT2: The good thing about recycling is that less energy is used than with new production. **(true)**
ACT3: Energy can be saved if one takes a shower instead of taking a bath. **(true)**

MULTIPLE-CHOICE QUESTIONS WITH ONE CORRECT ANSWER (IN BOLD)

ACT4: How can soil be protected from erosion?
by maintaining continuous vegetation
by letting the fields lie fallow
by plowing the fields regularly

ACT5: Which of the following statements is true? Asparagus from California is environmentally harmful because . . .
climatic conditions are not advantageous for growing asparagus in California.
too much packaging material is used.
air transport consumes excessive amounts of energy.

ACT6: In Germany, one of the following labels stands for certified organic cultivation. Which one?
"controlled organic"
"integrated agriculture"
"environmentally friendly farming"

ACT7: What is "gray energy"?
energy that was used for the production of an appliance
the total amount of energy used by an appliance
heat energy that is lost when appliances are used

ACT8: As a consequence of plowing fields . . .
the soil dries up.
the soil becomes compacted.
plants cannot absorb the humus.

ACT9: If ozone warnings are issued in the summertime, you should not drive . . .
because summer smog will be produced.
because otherwise, the hole in the ozone will increase.
due to the warm weather, the engine will give off more pollutants.

ACT10: What is the main cause of the increasing levels of nitrate pollution in groundwater?
more cars
agriculture
industrial air pollution
wastewater dumped in rivers

ACT11: Why is it better to collect and recycle aluminum than to throw it away?
because discarded aluminum gives off poisons when burned in incinerators
because producing new aluminum produces more poisonous materials than recycling does
because producing new aluminum consumes a large amount of energy

ACT12: Properly airing the house means . . .
leaving the window wide open for at least one hour.
airing briefly and powerfully with the heating turned off.
continuous ventilation with a tipped window.

ACT13: What is printed exclusively on recycled paper?
books
fashion magazines
newspapers

ACT14: In Germany, which certificate guarantees that paper was recycled?
the "Blue Angel"
"Aqua Pro Natura / World Park Tropical Forest"
the "Green Fir Tree"

ACT15: Which wood certificate guarantees sustainable forestry?
the "Rainforest Certificate"
the "FSC-Certificate"
the labeling "from licensed forestry"

ACT16: Where can someone dispose of old batteries?
in the residual waste
in the yellow bin
in the appropriate collection box in the supermarket

MULTIPLE-CHOICE QUESTIONS WITH MULTIPLE ANSWERS (IN BOLD)

ACT17: The energy consumption for heating can be reduced by . . .
keeping the room temperature constant.
setting the temperature lower at night.
insulating windows and doors.

ACT18: What can be done to save the (tropical) rain forests?
using recycled paper
abstaining from eating meat from South America
abstaining from eating meat from North America

ACT19: Using a personal computer can be made more environmentally friendly by . . .
turning the PC off when it is not being used for a longer time.
using a screensaver.
using a computer that is marked with an energy label.
always using recycled paper for printing.

ACT20: To counteract global warming, it makes sense to . . .
buy local food.
use public transportation instead of driving.
buy organic food.

ACT21: How can ozone buildup be reduced in the summertime?
by not using solvents
by not driving cars
by reducing the use of electricity

ACT22: In Germany, during which part of the year are which fruits or vegetables imported from other countries (or greenhouse produced)?
tomatoes in November
asparagus in May
peaches in April

ACT23: To keep water use as low as possible, you should water your garden . . .
in the morning.
at noon.
in the evening.

Effectiveness Knowledge

TRUE/FALSE QUESTIONS

EFF1: Incinerating waste is generally preferable to putting waste in a landfill. **(true)**

EFF2: For Italian-grown tomatoes, twice as much energy is used by the time they are sold in Germany as compared to locally grown tomatoes. **(true)**

EFF3: Comparing meat to vegetables (in amounts containing the same number of calories), the same amount of energy is needed. **(false)**

EFF4: It takes more energy to produce and transport batteries than the batteries themselves contain. **(true)**

EFF5: Per person and per kilometer, a car consumes 10 times more energy than a train. **(false)**

EFF6: Conventionally grown tomatoes consume only half the energy consumed by organically produced tomatoes. (**false**)

EFF7: It takes the same amount of energy to produce recycled paper as it takes to produce conventional paper. (**false**)

EFF8: A TV or stereo needs so little energy on "standby" that practically it makes no difference whether you turn it off completely. (**false**)

EFF9: Cooking 1.5 liters of soup needs 3 times more energy without a lid than with a lid. (**true**)

EFF10: Washing laundry at 60° Celsius reduces energy by 35% compared to at 90° Celsius. (**true**)

MULTIPLE-CHOICE QUESTIONS WITH ONE CORRECT ANSWER (IN BOLD)

EFF11: Recycling which of the following materials saves the most energy as compared to producing new material?
aluminum
glass
paper

EFF12: What type of milk packaging is more damaging to the environment?
paperboard cartons
returnable glass bottles

EFF13: What type of lamp consumes the least energy for the same amount of light?
conventional lightbulb
halogen lamp
fluorescent tube

EFF14: How much energy is required to grow wheat by integrated farming as compared to growing wheat by organic farming?
half the energy is required by integrated farming
the same amount is required
twice as much energy is required

EFF15: Returnable bottles can be reused up to . . .
10 times.
30 times.
60 times.

EFF16: Energy-saving lightbulbs consume . . .% less energy than conventional lightbulbs with the same illuminating power.
20%
80%

EFF17: What percentage of energy can be saved by using steamers instead of conventional cooking pots?
20%
50%
80%

EFF18: Water-saving showerheads consume . . . of the water consumed by conventional showerheads.
a quarter
half
three-quarters

EFF19: What has consumed the most energy up to the point at which Italian peppers are in the vegetable section of your grocery store?
heating the greenhouse
refrigerated storage
transport
packaging

EFF20: A household needs the most energy for . . .
lighting.
hot water.
heating.

EFF21: When cooking noodles, the most energy will be saved if . . .
water is cooked in a kettle.
less water is used.
warm tap water is heated.

EFF22: What is more environmentally friendly, exchanging components of an old PC or buying a new PC?
exchanging components of the old PC
buying a new PC

EFF23: How much water does it take to fill a bathtub?
50 liters
150 liters
300 liters

EFF24: For the production of an aluminum can . . . energy is used than for the production of a glass bottle.
twice as much
10 times more energy
20 times more energy

EFF25: How often can paper be reused by recycling?
6 to 7 times
4 to 5 times
2 to 3 times

EFF26: How many trees (the size of a spruce) are felled each year for one student?
3 trees
1 tree
2 trees

EFF27: Compared to a bus, a car emits . . . CO_2 per person.
twice as much
as much
more than 4 times as much
less

EFF28: Lowering the heating temperature at home by 1° Celsius means . . .% less energy consumption.
2%
4%
6%

EFF29: Each time a person goes to the toilet, . . . liters of drinking water disappear into the sewage system.
3 liters
12 liters
25 liters

Systems Thinking Scale (Davis and Stroink 2016, 577)

Note. This scale uses lower ratings for disagree and higher ratings for agreement; thus a higher score indicates greater systems thinking for statements without [r]. [r] denotes a reverse-keyed item (items that are NOT consistent with systems thinking). Thus disagreeing (scoring lower) with [r] statements would indicate higher levels of systems thinking. Reverse-keyed questions are used to prevent respondents from simply going through and clicking the same box each time. However, they can also confuse respondents; you may want to warn respondents to be very careful about their responses and to read each question carefully. (You may also want to change wording in reverse-keyed items if the wording is confusing for your respondents.)

Question. Please indicate your level of agreement with the following statements using the scale provided. There are no right or wrong answers.

Scale. 1–7 (strongly disagree—strongly agree)

1. When I have to make a decision in my life I tend to see all kinds of possible consequences to each choice.
2. Social problems, environmental problems, and economic problems are all separate issues. [r]
3. I like to know how events or information fit into the big picture.
4. Only very large events can significantly change big systems like economies or ecosystems. [r]
5. All the earth's systems, from the climate to the economy, are interconnected.
6. Everything is constantly changing.
7. Adding just one more small farm upstream from a lake can permanently alter that lake.
8. When a boom or a crash happens in part of the world's economy, it is because someone intentionally planned or designed for it to run that way. [r]
9. Ultimately, we can break all problems down to what is simply right and wrong. [r]
10. The earth, including all its inhabitants, is a living system.
11. Rules and laws should not change a lot over time. [r]
12. If I make plans and control my behavior I can accurately predict how my life will unfold. [r]
13. Seemingly small choices we make today can ultimately have major consequences.
14. My health has nothing to do with what is happening in the world. [r]
15. It is possible for a community to organize into a new form that was not planned or designed by an authority or government.

The Ultimate Cheatsheet for Critical Thinking (Global Digital Citizen Foundation, n.d.)

Note: This tool can be used as an embedded assessment. I suggest choosing questions and adapting them for your program as you see fit. You can then observe students' ability to discuss these questions.

Question. Want to exercise critical thinking skills? Ask these questions whenever you discover or discuss new information.

Scale. The questions are intended to spur critical thinking, and thus there is no scale.

Who . . .
benefits from this?
would be harmed by this?
makes decisions about this?
is most directly affected?
has also discussed this?
would be the best person to consult?
will be the key people in this?
deserves recognition for this?

What . . .
are the strengths/weaknesses?
is another perspective?
is another alternative?
would be a counterargument?
is the best/worst case scenario?
is most/least important?
can we do to make a positive change?
is getting in the way of our action?

Where . . .
would we see this in the real world?
are there similar concepts/situations?
is there the most need for this?
in the world would this be a problem?
can we get more information?
do we go for help with this?
will this idea take us?
are areas for improvement?

When . . .
is this acceptable/unacceptable?
would this benefit our society?
would this cause a problem?
is the best time to take action?
will we know we've succeeded?
has this played a part in our history?
can we expect this to change?
should we ask for help with this?

Why . . .
is this a problem/challenge?
is it relevant to me/others?
is this the best/worst scenario?
are people influenced by this?
should people know about this?
has it been this way for so long?
have we allowed this to happen?
is there a need for this today?

How . . .
is this similar to [something else]?
does this disrupt things?
do we know the truth about this?
will we approach this safely?
does this benefit us/others?
does this harm us/others?
do we see this in the future?
can we change this for our good?

Chapter 7. Values, Beliefs, and Attitudes

Values Scale (Steg, Perlaviciute, et al. 2014, 170)

Note. Whereas biospheric values are strongly associated with environmental behaviors, hedonic and other values are associated with environmental behaviors under certain circumstances. See chapter 7 for explanation.

Question. Rate the importance of these 16 values "as guiding principles in your lives." Vary your scores and rate only few values as extremely important.

Scale. 1–7 (opposed to my principles—extremely important)

Biospheric Values
BV1. Respecting the earth
BV2. Unity with nature
BV3. Protecting the environment
BV4. Preventing pollution

Altruistic values
AV1. Equality
AV2. A world at peace
AV3. Social justice
AV4. Being helpful

Egoistic values
EV1. Social power
EV2. Wealth
EV3. Authority
EV4. Influential
EV5. Ambitious

Hedonic values
HV1. Pleasure
HV2. Enjoying life
HV3. Gratification for oneself

Beliefs: New Ecological Paradigm (Dunlap et al. 2000, 433)

Note. I am including the New Ecological Paradigm because it is commonly used in environmental education and environmental behavior research. However, it has shown little ability to predict environmental behaviors. Thus I do ***not*** recommend implementing this survey if your goal is to determine how likely your program is to encourage environmental behaviors.

Question. Listed below are statements about the relationship between humans and the environment. For each one, please indicate your level of agreement or disagreement.

Scale. 1–5 (strongly agree—strongly disagree) (Agreement with the eight odd-numbered items and disagreement with the seven even-numbered items indicate pro-NEP responses.)

1. We are approaching the limit of the number of people the earth can support.
2. Humans have the right to modify the natural environment to suit their needs.
3. When humans interfere with nature it often produces disastrous consequences.
4. Human ingenuity will ensure that we do NOT make the earth unlivable.
5. Humans are severely abusing the environment.
6. The earth has plenty of natural resources if we just learn how to develop them.
7. Plants and animals have as much right as humans to exist.
8. The balance of nature is strong enough to cope with the impacts of modern industrial nations.
9. Despite our special abilities humans are still subject to the laws of nature.
10. The so-called "ecological crisis" facing humankind has been greatly exaggerated.
11. The earth is like a spaceship with very limited room and resources.
12. Humans were meant to rule over the rest of nature.
13. The balance of nature is very delicate and easily upset.
14. Humans will eventually learn enough about how nature works to be able to control it.
15. If things continue on their present course, we will soon experience a major ecological catastrophe.

Environmental Attitude Inventory (EAI) (Milfont and Duckitt 2010; Sutton and Gyuris 2015, 20–22)

Note. This is Sutton and Gyuris's (2015) shortened version of the original environmental attitudes inventory of Milfont and Duckitt (2010).

Question. Indicate your level of disagreement or agreement with the following statements.

Scale. 1–7 (strongly disagree—strongly agree); [r] indicates reverse coded items.

Scale 01. Enjoyment of nature

EN1. I really like going on trips into the countryside, for example to forests or fields.

EN2. I think spending time in nature is boring. [r]

EN3. Being out in nature is a great stress reducer for me.

Scale 02. Support for interventionist conservation policies

CP1. Industries should be able to use raw materials rather than recycled ones if this leads to lower prices and costs, even if it means the raw materials will eventually be used up. [r]

CP2. I am opposed to governments controlling and regulating the way raw materials are used to try and make them last longer. [r]

CP3. People in developed societies are going to have to adopt a more conserving lifestyle in the future.

Scale 03. Environmental movement activism

EA1. I would NOT get involved in an environmentalist organization. [r]

EA2. Environmental protection costs a lot of money. I am prepared to help out in a fund-raising effort.

EA3. I would not want to donate money to support an environmentalist cause. [r]

Scale 04. Conservation motivated by anthropocentric concern

AC1. Conservation is important even if it lowers peoples' standard of living. [r]

AC2. We need to keep rivers and lakes clean to protect the environment, and NOT as places for people to enjoy water sports. [r]

AC3. We should protect the environment even if it means people's welfare will suffer. [r]

Scale 05. Confidence in science and technology

CST1. Science and technology will eventually solve our problems with pollution, overpopulation, and diminishing resources.

CST2. The belief that advances in science and technology can solve our environmental problems is completely wrong and misguided. [r]

CST3. Modern science will solve our environmental problems.

Scale 06. Environmental fragility

EF1. People who say that the unrelenting exploitation of nature has driven us to the brink of ecological collapse are wrong. [r]

EF2. Humans are severely abusing the environment.

EF3. The idea that the balance of nature is terribly delicate and easily upset is much too pessimistic. [r]

Scale 07. Altering nature

AN1. I'd prefer a garden that is wild and natural to a well-groomed and ordered one. [r]

AN2. Human beings should not tamper with nature even when nature is uncomfortable and inconvenient for us. [r]

AN3. Turning new unused land over to cultivation and agricultural development should be stopped. [r]

AN4. When nature is uncomfortable and inconvenient for humans, we have every right to change and remake it to suit ourselves.

Scale 08. Personal conservation behavior

CB1. I am NOT the kind of person who makes efforts to conserve natural resources. [r]

CB2. Whenever possible, I try to save natural resources.

CB3. I always switch the light off when I don't need it on anymore.

Scale 09. Human dominance over nature

HD1. Humans are no more important than any other species. [r]

HD2. Human beings were created or evolved to dominate the rest of nature.

HD3. Plants and animals exist primarily to be used by humans.

Scale 10. Human utilization of nature

HU1. Protecting people's jobs is more important than protecting the environment.

HU2. Humans do NOT have the right to damage the environment just to get greater economic growth. [r]

HU3. The benefits of modern consumer products are more important than the pollution that results from their production and use.

Scale 11. Ecocentric concern

EC1. I do not believe protecting the environment is an important issue. [r]

EC2. Despite our special abilities, humans are still subject to the laws of nature.

EC3. It does NOT make me sad to see natural environments destroyed. [r]

Scale 12. Support for population growth policies

PG1. Families should be encouraged to limit themselves to two children or less.

PG2. We should never put limits on the number of children a couple can have. [r]

PG3. We would be better off if we dramatically reduced the number of people on earth.

Chapter 8. Nature Connectedness

Connectedness to Nature Scale (Mayer and Frantz 2004, 513)

Note. This scale has been used successfully to predict environmental behavior and feelings of well-being.

Question. Please answer each of these questions in terms of the way you generally feel.

Scale. 1–5 (strongly disagree—strongly agree); [r] indicates reverse coded items.

1. I often feel a sense of oneness with the natural world around me.
2. I think of the natural world as a community to which I belong.
3. I recognize and appreciate the intelligence of other living organisms.
4. I often feel disconnected from nature. [r]
5. When I think of my life, I imagine myself to be part of a larger cyclical process of living.
6. I often feel a kinship with animals and plants.
7. I feel as though I belong to the earth as equally as it belongs to me.
8. I have a deep understanding of how my actions affect the natural world.
9. I often feel part of the web of life.
10. I feel that all inhabitants of the earth, human, and nonhuman, share a common "life force."
11. Like a tree can be part of a forest, I feel embedded within the broader natural world.
12. When I think of my place on earth, I consider myself to be a top member of a hierarchy that exists in nature. [r]
13. I often feel like I am only a small part of the natural world around me, and that I am no more important than the grass on the ground or the birds in the trees.
14. My personal welfare is independent of the welfare of the natural world. [r]

Connection to Nature Scale for Children (Cheng and Monroe 2012, 41)

Note. This scale was developed for measuring connectedness to nature among children 9–10 years old. One item, "Being outdoors makes me happy," falls under enjoyment of nature and sense of oneness, but can be included just once when administering the survey.

Question. Indicate how much you agree with the following statements, with 1 being strongly disagree and 5 being strongly agree.

Scale. 1–5 (strongly disagree—strongly agree)

Enjoyment of nature

EN1. I like to hear different sounds in nature.

EN2. I like to see wild flowers in nature.

EN3. When I feel sad, I like to go outside and enjoy nature.

EN4. Being in the natural environment makes me feel peaceful.

EN5. I like to garden.

EN6. Collecting rocks and shells is fun.

EN7. Being outdoors makes me happy.

Empathy for creatures

EC1. I feel sad when wild animals are hurt.

EC2. I like to see wild animals living in a clean environment.

EC3. I enjoy touching animals and plants.

EC4. Taking care of animals is important to me.

Sense of oneness

SO1. Humans are part of the natural world.

SO2. People cannot live without plants and animals.

SO3. Being outdoors makes me happy.

Sense of responsibility

SR1. My actions will make the natural world different.

SR2. Picking up trash on the ground can help the environment.

SR3. People do not have the right to change the natural environment.

Nature Relatedness Scale (Short Version) (Nisbet and Zelenski 2013, 11)

Note. Nature relatedness is similar to nature connectedness. See table 8.1.

Question. Rate the extent to which you agree with each statement, using the scale from 1 to 5 as shown below. Please respond as you really feel, rather than how you think "most people" feel.

Scale. 1–5 (strongly disagree—strongly agree)

1. My ideal vacation spot would be a remote, wilderness area.
2. I always think about how my actions affect the environment.
3. My connection to nature and the environment is a part of my spirituality.
4. I take notice of wildlife wherever I am.
5. My relationship to nature is an important part of who I am.
6. I feel very connected to all living things and the earth.

Chapter 9. Sense of Place

Sense of Place Scale (Kudryavtsev, Krasny, and Stedman 2012, 4)

Note. This scale was used in a study of the impact of environmental education programs on youth place attachment and ecological place meaning in the Bronx. You can adapt the questions for other places and programs.

Question. Indicate your level of agreement or disagreement with the following statements.

Scale. 1–5 (strongly disagree—strongly agree); [r] indicates reverse coded items.

Place attachment

PA 1. The Bronx is the best place for what I like to do.
PA 2. I feel like the Bronx is part of me.
PA 3. Everything about the Bronx reflects who I am.
PA 4. I am more satisfied in the Bronx than in other places.
PA 5. I identify myself strongly with the Bronx.
PA 6. The Bronx is not a good place for what I enjoy doing. [r]
PA 7. There are better places to be than the Bronx. [r]
PA 8. The Bronx reflects the type of person I am.

Ecological place meaning

The Bronx is a place:
PM 1. to connect with nature.
PM 2. to watch animals and birds.
PM 3. where people can find nature.
PM 4. where trees are an important part of community.
PM 5. where people have access to rivers.
PM 6. where people come to community gardens.
PM 7. where people have access to parks.
PM 8. to canoe and boat.
PM 9. to have fun in nature.
PM 10. to learn about nature.
PM 11. to enjoy nature's beauty.
PM 12. to grow food.

Chapter 10. Efficacy

Self-Efficacy and Collective Efficacy (Reese and Junge 2017, 6)

Note. Self- and collective efficacy scales include two general environmental protection items and one specific plastics-related item. You can adapt questions for different groups and goals.

Question. Indicate level of agreement with following statements.

Scale. 1–7 (not true at all—completely true)

SELF-EFFICACY

1. I am optimistic that I can protect the environment.
2. I am capable of protecting the environment.
3. I think that I am capable of protecting the environment by means of my personal plastic reduction.

COLLECTIVE EFFICACY

1. I am optimistic that we as plastic challenge participants can protect the environment together.
2. We as plastic challenge participants have the capability to protect the environment.
3. I think we as plastic challenge participants can collectively protect the environment with reducing plastic usage.

Group, Participative, and Individual Efficacy (Van Zomeren, Saguy, and Schellhaas 2013, 16–17)

Note. Fill in words in parentheses depending on the study context, including relevant group (e.g., students in our school) and goal (e.g., can persuade city to install solar power). Feel free to simply use the questions marked by an asterisk if you prefer a shorter survey.

Question. In this survey we are interested in your opinion about (relevant goal).

Scale. 1–7 (not at all—very much)

GROUP EFFICACY BELIEFS

1. I believe that (. . .), as a group, can (. . .).
2. I believe that (. . .), together, can (. . .).*
3. I believe that (. . .), through joint actions, can (. . .).*
4. I believe that (. . .) can achieve their common goal of (. . .).

PARTICIPATIVE EFFICACY BELIEFS

1. I believe that I, as an individual, can contribute greatly so that (. . .), as a group, can (. . .).
2. I believe that I, as an individual, can provide an important contribution so that (. . .), together, can (. . .).*
3. I believe that I, as an individual, can provide a significant contribution so that, through joint actions, (. . .) can (. . .).*
4. I believe that I, as an individual, can contribute meaningfully so that (. . .) can achieve their common goal of (. . .).

INDIVIDUAL EFFICACY BELIEFS

1. I believe that I, as an individual, can (. . .).*
2. I believe that I can (. . .).
3. I believe that I, through individual actions, can (. . .).*
4. I believe that I can achieve my personal goal of (. . .).

Political Efficacy (Schulz and Sibberns 2004, 258)

Note. These questions are part of a larger survey used to measure civic education outcomes among youth across multiple countries (Schulz 2005). Questions can be adapted as "school efficacy" (e.g., "Our school cares a lot about what all of us think about new school policies"). You can also adapt questions for specific environmental issues (e.g., "When global warming is discussed, I have something to say").

Question. How do you feel about the following statements?

Scale. 1–4 (strongly disagree—strongly agree)

INTERNAL POLITICAL EFFICACY

1. I know more about politics than most people my age.
2. When political issues or problems are being discussed, I usually have something to say.
3. I am able to understand most political issues easily.

EXTERNAL POLITICAL EFFICACY

1. The government cares a lot about what all of us think about new laws.
2. The government is doing its best to find out what people [ordinary people] want.
3. When people get together [organize] to demand change, the leaders in government listen.

School Efficacy (Torney-Purta et al. 2001, 206)

Note. This survey was developed for youth across multiple countries and may be more appropriate than political efficacy in contexts where youth have limited opportunity to participate in political life.

Question. How do you feel about the following statements?

Scale. 1–4 (strongly disagree—strongly agree)

1. Electing student representatives to suggest changes in how the school is run makes schools better.
2. Lots of positive changes happen in this school when students work together.
3. Organizing groups of students to state their opinions could help solve problems in this school.
4. Students acting together can have more influence on what happens in this school than students acting alone.

Civic Efficacy (Syvertsen, Wray-Lake, and Metzger 2015, 11)

Note. This survey was developed for use with youth. You may want to revise or add questions to reflect civic efficacy related to environmental issues.

Question. How much do you disagree or agree with each statement?

Scale. 1–5 (strongly disagree—strongly agree)

1. I can make a positive difference in my community.
2. Even though I am a teenager, there are ways for me to get involved in my community.
3. I can use what I know to solve "real life" problems in my community.

Chapter 11. Identity

Environmental Identity Scale (Clayton 2003)

Note. Scale available for downloading at http://discover.wooster.edu/sclayton/files/2011/11/env-id-memo.pdf.

Question. Please indicate the extent to which each of the following statements describes you by using the appropriate number from the scale below.

Scale. 1–7 (not at all true—completely true)

1. I spend a lot of time in natural settings (woods, mountains, desert, lakes, ocean).
2. Engaging in environmental behaviors is important to me.
3. I think of myself as a part of nature, not separate from it.
4. If I had enough time or money, I would certainly devote some of it to working for environmental causes.
5. When I am upset or stressed, I can feel better by spending some time outdoors "communing with nature."
6. Living near wildlife is important to me; I would not want to live in a city all the time.
7. I have a lot in common with environmentalists as a group.
8. I believe that some of today's social problems could be cured by returning to a more rural lifestyle in which people live in harmony with the land.
9. I feel that I have a lot in common with other species.
10. I like to garden.
11. Being a part of the ecosystem is an important part of who I am.
12. I feel that I have roots to a particular geographical location that had a significant impact on my development.
13. Behaving responsibly toward the earth—living a sustainable lifestyle—is part of my moral code.
14. Learning about the natural world should be an important part of every child's upbringing.
15. In general, being part of the natural world is an important part of my self image.
16. I would rather live in a small room or house with a nice view than a bigger room or house with a view of other buildings.
17. I really enjoy camping and hiking outdoors.
18. Sometimes I feel like parts of nature—certain trees, or storms, or mountains—have a personality of their own.
19. I would feel that an important part of my life was missing if I was not able to get out and enjoy nature from time to time.

20. I take pride in the fact that I could survive outdoors on my own for a few days.
21. I have never seen a work of art that is as beautiful as a work of nature, like a sunset or a mountain range.
22. My own interests usually seem to coincide with the position advocated by environmentalists.
23. I feel that I receive spiritual sustenance from experiences with nature.
24. I keep mementos from the outdoors in my room, like shells or rocks or feathers.

Chapter 12. Norms

Injunctive Social Norms (Terry, Hogg, and White 1999, 232)

Note: Injunctive norms are also referred to as "perceived social norms."

Scale: question 1: 1–7 (strongly approve—strongly disapprove); question 2: 1–7 (should—should not)

1. If I engaged in household recycling during the next two weeks, most people who are important to me would . . .
2. Most people who are important to me think that I should/shouldn't engage in household recycling during the next two weeks.

Descriptive Social Norms (Nigbur, Lyon, and Uzzell 2010; Nolan et al. 2008, 916)

Note. You can replace recycling or conserving energy with other behaviors and the group with more relevant groups (e.g., residents of your city). Although the questions use different scales, feel free to adjust the scales so they are consistent if you administer all three questions. You may also be able to obtain real information on the actual descriptive social norm—for example, how many people in a neighborhood use solar power.

Scale. Questions 1 and 2: 1–7 (none, a few, some, around half, many, most, and all); question 3: 1–4 (never—almost always)

1. Give a rough estimate of the proportion of households in the neighborhood that participated in Green Box recycling.
2. How many of your friends and peers would engage in household recycling?
3. How often do you think residents of your state try to conserve energy?

Personal Norms (Van der Werff, Steg, and Keizer 2013, 10; questions adapted from the original)

Note. The first three questions are general; the last three specific questions can be adapted for other environmental behaviors.

Question. Indicate level of agreement with the following statements.

Scale. 1–7 (totally disagree—totally agree)

1. I feel morally obliged to act in an environmentally friendly manner.
2. I would feel guilty if I did not act in an environmentally friendly manner.

3. Acting environmentally friendly would give me a good feeling.
4. I feel morally obliged to bring a reusable cup to school.
5. I would feel guilty if I bought plastic straws.
6. I feel proud when I join a tree planting volunteer activity.

Chapter 13. Social Capital

Social Capital Survey for Youth (Krasny, Kalbacker, et al. 2013, 12–13)

Note. This survey includes questions on aspects of social capital including social trust and informal socializing. Items are adapted from the National Social Capital Benchmark Survey (Roper Center for Public Opinion Research 2000) and social capital evaluation guidelines for nonprofit organizations (Kennedy School of Government 2012).

SOCIAL TRUST

Question. Please indicate the level of your agreement with these statements about your relationships with other people in your community.

Scale. 1–5 (strongly disagree—strongly agree); [r] indicates reverse coded items.

ST1. In general, I can trust most people.
ST2. I do not trust people in my neighborhood. [r]
ST3. I trust people I go to school with.
ST4. I trust people I hang out with.
ST5. I do not trust the police in my neighborhood. [r]

INFORMAL SOCIALIZING

Question. How often do you . . .

Scale. 1–5 (never—very often)

IS1. have friends over to your home?
IS2. attend a celebration, parade, or art event in your community?
IS3. attend a local sports event in your community?
IS4. visit relatives in person or have them come visit you?
IS5. hang out with friends at a park, shopping mall, or other public place?

DIVERSITY OF FRIENDSHIP

Question. Please check all that apply to you.

Scale. 1 = yes, 0 = no

DF1. I have close friends that are all ages, not just my age.
DF2. I have close friends who are other races than me.
DF3. I have close friends who have other favorite interests than me.
DF4. I have other close friends who go to other schools than me.

DF5. I have close friends who are from other countries.

DF6. I have close friends whose families have more money or less money than my family.

ASSOCIATIONAL INVOLVEMENT

Note. We suggest changing these items to those that might change as a result of an environmental education program.

Question. In what kinds of education programs do you currently participate? Check all that apply to you.

Scale. Question uses checkbox responses.

AI1. An after-school program

AI2. A youth club such as a Boys and Girls Club, Scouts, or a 4-H club

AI3. A community service club

AI4. A band, orchestra, or choir

AI5. A sports team

AI6. Another club or organization

CIVIC LEADERSHIP

Note. We suggest changing these items to those that might change as a result of an environmental education program.

Question. Check all that apply to you.

Scale. Question uses checkbox responses.

CL1. I am on student council or student government.

CL2. I am on a planning team for a school organization.

CL3. I am a class officer.

CL4. I am an officer of a club.

CL5. I am a team captain of a sports team.

Chapter 14. Positive Youth Development

Note. Positive youth development encompasses many outcomes, as suggested by Lerner's six C's: competence, confidence, connection, character, caring, and contribution (Lerner et al. 2005). I have chosen to focus on contribution, or active citizenship, as particularly relevant to environmental education, including programs that have a social justice focus. Some questions overlap with those included in environmental action (chapter 5), political efficacy (chapter 10), and social capital (chapter 13). You may want to adapt questions to be more relevant for your program and audiences. For a comprehensive set of positive youth development survey instruments used across multiple countries see Hinson et al. (2016).

Active and Engaged Citizenship (Zaff et al. 2010, 743)

CIVIC DUTY

Note. Within civic duty, scales vary. I have grouped questions using same scale. [r] indicates reverse coded items.

Question. How important are the following to you?
Scale. 1–5 (not important—extremely important)

CD1. Helping to reduce hunger and poverty in the world
CD2. Helping to make sure all people are treated fairly
CD3. Helping to make the world a better place to live in
CD4. Helping other people
CD5. Speaking up for equality (everyone should have the same rights and opportunities)

Question. Rate your level of agreement with the following statements.
Scale. 1–5 (strongly disagree—strongly agree)

CD6. It's not really my problem if my neighbors are in trouble and need help. [r]
CD7. I believe I can make a difference in my community.
CD8. I often think about doing things so that people in the future can have things better.
CD9. It is important to me to contribute to my community and society.

Question. How well do the following statements describe you?
Scale. 1–5 (not very well—very well)

CD10. When I see someone being taken advantage of, I want to help them.
CD11. When I see someone being treated unfairly, I don't feel sorry for them. [r]
CD12. I feel sorry for other people who don't have what I have.

CIVIC SKILLS

Question. Indicate the extent to which you can:
Scale. 1–5 (Definitely can't—definitely can)

CS1. Contact a newspaper, radio, or TV talk show to express your opinion on an issue
CS2. Contact an elected official about the problem
CS3. Contact or visit someone in government who represents your community
CS4. Write an opinion letter to a local newspaper
CS5. Express your views in front of a group of people
CS6. Sign an e-mail or written petition

NEIGHBORHOOD SOCIAL CONNECTION

Question. Indicate the degree to which you disagree or agree with following items.
Scale. 1–5 (strongly disagree—strongly agree)

NS1. Adults in my town or city listen to what I have to say.
NS2. Adults in my town or city make me feel important.
NS3. In my town or city, I feel like I matter to people.
NS4. In my neighborhood, there are lots of people who care about me.
NS5. If one of my neighbors saw me do something wrong, he or she would tell one of my parents.
NS6. My teachers really care about me.

CIVIC PARTICIPATION

Question. How often do you . . . ?
Scale. Scales vary depending on question. Same-scale items grouped together.
1–5 (never—very often)

CP1. Help make your city or town a better place for people to live
CP2. Help out at your church, synagogue, or other place of worship
CP3. Help a neighbor
CP4. Help out at your school

Scale. 1–6 (never—everyday)

CP5. Volunteer your time (at a hospital, day care center, food bank, youth program, community service agency)
CP6. Mentor / offer peer advice
CP7. Tutor

Scale. 1–6 (never—5 or more times)

CP8. During the last 12 months, how many times have you been a leader in a group or organization?

Chapter 15. Health and Well-Being

We include scales for well-being and climate change hope and coping.

Well-Being Scale (McLellan and Steward 2015, 316)

Note. The survey was developed for use in schools but could be used in other settings.

Question. Indicate how true the following statements are for you.

Scale. 1–5 (never true—always true); [r] indicates reverse coded items.

1. I feel good about myself.
2. I feel healthy.
3. I feel I am doing well.
4. I feel miserable. [r]
5. I feel I have lots of energy.
6. I feel cared for.
7. I feel valuable.
8. I feel worried. [r]
9. I feel I can deal with problems.
10. I feel bored. [r]
11. I feel noticed.
12. I feel people are friendly.
13. I feel there is lots to look forward to.
14. I feel safe.
15. I feel confident.
16. I feel a lot of things are a real effort. [r]
17. I feel I enjoy things.
18. I feel lonely. [r]
19. I feel excited by lots of things.
20. I feel happy.
21. I feel I'm treated fairly.

Climate Change Hope Scale (Li and Monroe 2017a, 470)

Note. Developed for use with secondary school students.

Question. State the extent to which you agree or disagree with the following statements.

Scale. 1–7 (strongly agree—strongly disagree); [r] indicates reverse coded items.

PERSONAL-SPHERE WILL AND WAY (PW)

PW1. I am willing to take actions to help solve problems caused by climate change.

PW2. I know that there are things that I can do to help solve problems caused by climate change.

PW3. I know what to do to help solve problems caused by climate change.

PW4. At the present time, I am energetically pursuing ways to solve problems caused by climate change.

COLLECTIVE-SPHERE WILL AND WAY (CW)

CW1. If everyone works together, we can solve problems caused by climate change.

CW2. I believe that scientists will be able to find ways to solve problems caused by climate change.

CW3. 1. I believe people will be able to solve problems caused by climate change.

CW4. I believe more people are willing to take actions to help solve problems caused by climate change.

CW5. Even when some people give up, I know there will be others who will continue to try to solve problems caused by climate change.

CW6. Every day, more people begin to care about problems caused by climate change.

CW7. Because people can learn from their mistakes, they will eventually mitigate and adapt to climate change.

LACK OF WILL AND WAY (LW)

LW1. Climate change is beyond my control, so I won't even bother trying to solve problems caused by climate change. [r]

LW2. The actions I can take are too small to help solve problems caused by climate change. [r]

LW3. Climate change is so complex we will not be able to solve problems that it causes. [r]

LW4. I can't think of what I can do to help solve problems caused by climate change. [r]

Climate Change Coping Scale (Ojala 2012b, 229)

Note. Developed for use with children/youth. Emotion-focused coping is generally considered a negative factor.

Question. When one hears about societal problems such as climate change, one can feel worried or upset. Below is a list, and for every item we would like you to indicate how well it applies to what you do or think when you are reminded of climate change. Choose the alternative that you feel best applies to you, and choose only one alternative per item.

Scale. 1–5 (not true at all—completely true)

MEANING-FOCUSED COPING

MF1. More and more people have started to take climate change seriously.
MF2. I have faith in humanity; we can fix all problems.
MF3. I trust scientists to come up with a solution in the future.
MF4. I have faith in people engaged in environmental organizations.
MF5. I trust the politicians.
MF6. Even though it is a big problem, one has to have hope.

EMOTION-FOCUSED COPING (DE-EMPHASIZING / DON'T CARE)

EF1. I think that the problem is exaggerated.
EF2. I don't care since I don't know much about climate change.
EF3. Climate change is something positive because the summers will get warmer.
EF4. I can't be bothered to care about climate change.
EF5. Nothing serious will happen during my lifetime.
EF6. Climate change does not concern those of us living in [name country].

PROBLEM-FOCUSED COPING

PF1. I think about what I myself can do.
PF2. I search for information about what I as a child can do.
PF3. I talk with my family and friends about what one can do to help.

References

Abhyankar, A., and M. E. Krasny. 2018. "From Practice to Fledging Social Movement in India: Lessons from 'The Ugly Indian.'" In Krasny, *Grassroots to Global*, 231–248.

Abigail, J. 2016. "Cultivating Mangroves." In *Civic Ecology: Stories about Love of Life, Love of Place*, edited by M. E. Krasny and K. Snyder, 1–5. Ithaca, NY: Cornell University Civic Ecology Lab.

ACGA. 2018. American Community Gardening Association. http://www.community garden.org/.

Adams, J. D., D. A. Greenwood, M. Thomashow, and A. Russ. 2017. "Sense of Place." In Russ and Krasny, *Urban Environmental Education Review*, 68–75.

Aguilar, O., A. Price, and M. E. Krasny. 2015. "Perspectives on Community Environmental Education." In *Across the Spectrum: Resources for Environmental Educators*, edited by M. Monroe and M. E. Krasny, 169–183. Washington, DC: NAAEE.

Ahn, T. K., and E. Ostrom, eds. 2003. *Foundations of Social Capital*. Cheltenham, UK: Edward Elgar.

——. 2008. "Social Capital and Collective Action." In *Handbook of Social Capital*, edited by D. Castiglione, J. W. van Deth, and G. Wolleb, 70–100. Oxford: Oxford University Press.

Ajzen, I. 1991. "The Theory of Planned Behavior." *Organizational Behavior and Human Decision Processes* 50(2): 179–211.

Alisat, S., and M. Riemer. 2015. "The Environmental Action Scale: Development and Psychometric Evaluation." *Journal of Environmental Psychology* 43: 13–23.

Altman, I., and S. Low. 1994. "Place Attachment." In *Place Attachment*, edited by I. Altman and S. Low, 1–12. New York: Plenum.

AMNH. 2012. "Great Pollinator Project." http://greatpollinatorproject.org/.

Amsden, B. L., R. C. Stedman, and L. E. Kruger. 2010. "The Creation and Maintenance of Sense of Place in a Tourism-Dependent Community." *Leisure Sciences* 33(1): 32–51.

Andersson, E., S. Barthel, and K. Ahrne. 2007. "Measuring Social-Ecological Dynamics behind the Generation of Ecosystem Services." *Ecological Applications* 17(5): 1267–1278.

Ardoin, N. M. 2006. "Toward an Interdisciplinary Understanding of Place: Lessons for Environmental Education." *Canadian Journal of Environmental Education* 11: 14.

——. 2014. "Exploring Sense of Place and Environmental Behavior at an Ecoregional Scale in Three Sites." *Human Ecology* 42(3): 425–441.

Ardoin, N. M., K. Biedenweg, and K. O'Connor. 2015. "Evaluation in Residential Environmental Education: An Applied Literature Review of Intermediary Outcomes." *Applied Environmental Education & Communication* 14(1): 43–56.

Ardoin, N. M., M. DiGiano, J. Bundy, S. Chang, N. Holthuis, and K. O'Connor. 2014. "Using Digital Photography and Journaling in Evaluation of Field-Based Environmental Education Programs." *Studies in Educational Evaluation* 41: 68–76.

Ardoin, N. M., M. DiGiano, K. O'Connor, and N. Holthuis. 2016. "Using Online Narratives to Explore Participant Experiences in a Residential Environmental Education Program." *Children's Geographies* 14(3): 263–281.

Ardoin, N. M., M. DiGiano, K. O'Connor, and T. Podkul. 2016. "The Development of Trust in Residential Environmental Education Programs." *Environmental Education Research* 23(9): 1335–1355.

Armitage, D., M. Marschke, and R. Plummer. 2008. "Adaptive Co-management and the Paradox of Learning." *Global Environmental Change* 18: 86–98.

Armitage, D., R. Plummer, F. Berkes, R. I. Arthur, A. T. Charles, I. J. Davidson-Hunt, A. P. Diduck, et al. 2009. "Adaptive Management for Social-Ecological Complexity." *Frontiers in Ecology and the Environment* 7(2): 95–102.

Armstrong, A., M. E. Krasny, and J. Schuldt. 2018. *Climate Change Communication for Environmental Educators*. Ithaca, NY: Cornell University Press.

Ask Umbra. 2016. "What's Worse, Burning Plastic or Sending It to a Landfill?" Grist. https://grist.org/living/whats-worse-burning-plastic-or-sending-it-to-a-landfill/.

Assaraf, O. B. Z., and N. Orion. 2005. "Development of System Thinking Skills in the Context of Earth System Education." *Journal of Research in Science Teaching* 42(5): 518–560.

Atkins, R., and D. Hart. 2003. "Neighborhoods, Adults, and the Development of Civic Identity in Urban Youth." *Applied Developmental Science* 7(3): 156–164.

Bakker, T. P., and C. H. de Vreese. 2011. "Good News for the Future? Young People, Internet Use, and Political Participation." *Communication Research* 38(4): 451–470.

Ballantyne, R., S. Connell, and J. Fien. 2006. "Students as Catalysts of Environmental Change: A Framework for Researching Intergenerational Influence through Environmental Education." *Environmental Education Research* 12(3): 413–427.

Ballew, M., A. Omoto, and P. Winter. 2015. "Using Web 2.0 and Social Media Technologies to Foster Proenvironmental Action." *Sustainability* 7(8): 10620.

Bamberg, S. 2003. "How Does Environmental Concern Influence Specific Environmentally Related Behaviors? A New Answer to an Old Question." *Journal of Environmental Psychology* 23(1): 21–32.

Bamberg, S., and G. Möser. 2007. "Twenty Years after Hines, Hungerford, and Tomera: A New Meta-analysis of Psycho-Social Determinants of Pro-environmental Behaviour." *Journal of Environmental Psychology* 27(1): 14–25.

Bandura, A. 1977. "Self-Efficacy: Toward a Unifying Theory of Behavioral Change." *Psychological Review* 84(2): 191–215.

——. 1993. "Perceived Self-Efficacy in Cognitive Development and Functioning." *Educational Psychologist* 28(2): 117–148.

——. 1997. *Self-Efficacy: The Exercise of Control*. New York: W. H. Freeman, 604.

——. 2004. "Social Cognitive Theory for Personal and Social Change by Enabling Media." In *Entertainment-Education and Social Change: History, Research, and Practice*, edited by A. Singhal, M. J. Cody, E. M. Rogers, and M. Sabido, 75–96. Mahwah, NJ: Lawrence Erlbaum.

Barrett, M., and I. Brunton-Smith. 2014. "Political and Civic Engagement and Participation: Towards an Integrative Perspective." *Journal of Civil Society* 10(1): 5–28.

Barth, M., P. Jugert, and I. Fritsche. 2016. "Still Underdetected—Social Norms and Collective Efficacy Predict the Acceptance of Electric Vehicles in Germany." *Transportation Research Part F: Traffic Psychology and Behaviour* 37: 64–77.

Barthel, S., S. Belton, M. Giusti, and C. M. Raymond. 2018. "Fostering Children's Connection to Nature through Authentic Situations: The Case of Saving Salamanders at School." *Frontiers in Psychology* 9: 928.

Beaumont, E. 2010. "Political Agency and Empowerment: Pathways for Developing a Sense of Political Efficacy in Young Adults." In *Handbook of Research on Civic Engagement in Youth*, edited by L. R. Sherrod, J. Torney-Purta, and C. Flanagan, 525–558. Somerset, NJ: John Wiley & Sons.

Beery, T. H., and D. Wolf-Watz. 2014. "Nature to Place: Rethinking the Environmental Connectedness Perspective." *Journal of Environmental Psychology* 40: 198–205.

Beilin, R., and A. Hunter. 2011. "Co-constructing the Sustainable City: How Indicators Help Us 'Grow' More Than Just Food in Community Gardens." *Local Environment* 16(6): 523–538.

Belluigi, D. Z., and G. Cundill. 2017. "Establishing Enabling Conditions to Develop Critical Thinking Skills: A Case of Innovative Curriculum Design in Environmental Science." *Environmental Education Research* 23(7): 950–971.

Bem, D. J. 1972. "Self-Perception Theory." In *Advances in Experimental Social Psychology*, vol. 6, edited by L. Berkowitz, 1–62. Amsterdam: Elsevier.

Bennett, W. L., and A. Segerberg. 2013. *The Logic of Connective Action: Digital Media and the Personalization of Contentious Politics*. New York: Cambridge University Press.

Berkes, F. 2004. "Knowledge, Learning and the Resilience of Social-Ecological Systems. Knowledge for the Development of Adaptive Co-management." Tenth Biennial Conference of the International Association for the Study of Common Property, Oaxaca, Mexico.

Berkes, F., J. Colding, and C. Folke. 2003. *Navigating Social-Ecological Systems: Building Resilience for Complexity and Change*. Cambridge: Cambridge University Press.

Bernstein, J. M. 2017. "Renewing the New Environmental Paradigm Scale: The Underlying Diversity of Contemporary Environmental Worldviews." PhD diss., University of Hawai'i.

Bettertogether. 2000. "Youth and Social Capital." Cambridge, MA: Saguaro Seminar on Civic Engagement in America.

Biggs, D., N. Abel, A. T. Knight, A. Leitch, A. Langston, and N. C. Ban. 2011. "The Implementation Crisis in Conservation Planning: Could 'Mental Models' Help?" *Conservation Letters* 4(3): 169–183.

Biggs, R., M. Schlüter, D. Biggs, E. L. Bohensky, S. BurnSilver, G. Cundill, V. Dakos, et al. 2012. "Toward Principles for Enhancing the Resilience of Ecosystem Services." *Annual Review of Environment and Resources* 37(1): 421–448.

Biggs, R., M. Schlüter, and M. L. Schoon. 2015. "An Introduction to the Resilience Approach and Principles to Sustain Ecosystem Services in Social-Ecological Systems." In *Principles for Building Resilience: Sustaining Ecosystem Services in Social-Ecological Systems*, edited by M. Schlüter, M. L. Schoon, and R. Biggs, 1–31. Cambridge: Cambridge University Press.

Binghamton (New York). 2016. "City Secures $2.7 Million FEMA Demolition Grant." http://www.binghamton-ny.gov/city-secures-27-million-fema-demolition-grant.

Blair, D. 2009. "The Child in the Garden: An Evaluative Review of the Benefits of School Gardening." *Journal of Environmental Education* 40(2): 15–38.

Blatt, E. N. 2013. "Exploring Environmental Identity and Behavioral Change in an Environmental Science Course." *Cultural Studies of Science Education* 8: 466–488.

Bonanno, G. A. 2004. "Loss, Trauma, and Human Resilience: How We Have Underestimated the Human Capacity to Thrive after Extremely Aversive Events." *American Psychologist* 59(1): 20–28.

Bourdieu, P. 1986. "The Forms of Capital." In *Handbook of Theory and Research for the Sociology of Education*, edited by J. G. Richardson, 241–258. New York: Greenwood.

Bowers, C. A. 2002. "Toward an Eco-justice Pedagogy." *Environmental Education Research* 8(1): 21–34.
Brechin, S. R., and W. Kempton. 1994. "Global Environmentalism: A Challenge to the Postmaterialism Thesis?" *Social Science Quarterly* 75(2): 245–269.
Briggs, L., M. E. Krasny, and R. C. Stedman. 2019. "Exploring Youth Development through an Environmental Education Program for Rural Indigenous Women." *Journal of Environmental Education* 50(1): 37–51.
Briggs, L., R. Stedman, and M. E. Krasny. 2014. "Photo-Elicitation Methods in Studies of Children's Sense of Place." *Children, Youth and Environments* 24(3): 153–172.
Brown, B., D. D. Perkins, and G. Brown. 2003. "Place Attachment in a Revitalizing Neighborhood: Individual and Block Levels of Analysis." *Journal of Environmental Psychology* 23(3): 259–271.
Brügger, A., F. G. Kaiser, and N. Roczen. 2011. "One for All? Connectedness to Nature, Inclusion of Nature, Environmental Identity, and Implicit Association with Nature." *European Psychologist* 16(4): 324–333.
Brundtland Commission. 1987. *Our Common Future*. Oxford: Oxford University Press.
Burbaugh, B., M. Siebel, and T. Archibald. 2017. "Using a Participatory Approach to Investigate a Leadership Program's Theory of Change." *Journal of Leadership Education* online.
Bureau of Labor Statistics. 2016. "Volunteering in the United States, 2015." https://www.bls.gov/news.release/volun.nr0.htm.
C&NN. 2017. Children & Nature Network. http://www.childrenandnature.org/.
Canada (Government of). 2017. "Canada's Food Guide Guiding Principles." https://www.foodguideconsultation.ca/guiding-principles-detailed.
Capaldi, C. A., R. L. Dopko, and J. M. Zelenski. 2014. "The Relationship between Nature Connectedness and Happiness: A Meta-analysis." *Frontiers in Psychology* 5: 976.
Capra, F. 2007. "Sustainable Living, Ecological Literacy, and the Breath of Life." *Canadian Journal of Education* 12: 9–18.
Capra, F., and P. L. Luisi. 2014. *The Systems View of Life: A Unifying Vision*. Cambridge: Cambridge University Press.
Carleton-Hug, A., and J. W. Hug. 2010. "Challenges and Opportunities for Evaluating Environmental Education Programs." *Evaluation and Program Planning* 33(2): 159–164.
CARRI. 2013. "Definitions of Community Resilience: An Analysis." Washington, DC: Community and Regional Resilience Institute, 14.
Carrus, G., M. Bonaiuto, and M. Bonnes. 2005. "Environmental Concern, Regional Identity, and Support for Protected Areas in Italy." *Environment and Behavior* 37(2): 237–257.
Catalano, R. F., M. L. Berglund, A. M. R. Jean, S. L. Heather, and J. D. Hawkins. 2004. "Positive Youth Development in the United States: Research Findings on Evaluations of Positive Youth Development Programs." *Annals of the American Academy of Political and Social Science* 591: 98–124.
Cermak, M. J. 2012. "Hip-Hop, Social Justice, and Environmental Education: Toward a Critical Ecological Literacy." *Journal of Environmental Education* 43(3): 192–203.
Cervinka, R., K. Röderer, and E. Hefler. 2011. "Are Nature Lovers Happy? On Various Indicators of Well-Being and Connectedness with Nature." *Journal of Health Psychology* 17(3): 379–388.
Chao, Y.-L., and S.-P. Lam. 2011. "Measuring Responsible Environmental Behavior: Self-Reported and Other-Reported Measures and Their Differences in Testing a Behavioral Model." *Environment and Behavior* 43(1): 53–71.

Chavis, D. M., and A. Wandersman. 1990. "Sense of Community in the Urban Environment: A Catalyst for Participation and Community Development." *American Journal of Community Psychology* 18(1): 55–81.

Chawla, L. 1994. "Childhood Place Attachments." In *Place Attachment*, edited by I. Altman and S. Low, 63–86. New York: Plenum.

——. 2001. "Putting Young Old Ideas into Action: The Relevance of *Growing Up in Cities* to Local Agenda 21." *Local Environment* 6(1): 13–25.

——. 2006. "Learning to Love the Natural World Enough to Protect It." *Barn* 2: 57–78.

——. 2009. "Growing Up Green: Becoming an Agent of Care for the Natural World." *Journal of Developmental Processes* 41(1): 6–23.

——. 2015. "Benefits of Nature Contact for Children." *Journal of Planning Literature* 30(4): 433–452.

Chawla, L., and D. F. Cushing. 2007. "Education for Strategic Environmental Behavior." *Environmental Education Research* 13(4): 437–452.

Chawla, L., and V. Derr. 2012. "Developing Conservation Behaviors in Childhood and Youth." In *The Oxford Handbook of Environmental and Conservation Psychology*, edited by S. Clayton, 527–555. New York: Oxford University Press.

Checkoway, B. 2012. "Education for Democracy by Young People in Community-Based Organizations." *Youth & Society* 45(3): 389–403.

Checkoway, B., and A. Aldana. 2013. "Four Forms of Youth Civic Engagement for Diverse Democracy." *Children and Youth Services Review* 35(11): 1894–1899.

Chen, M.-F. 2015. "Self-Efficacy or Collective Efficacy within the Cognitive Theory of Stress Model: Which More Effectively Explains People's Self-Reported Proenvironmental Behavior?" *Journal of Environmental Psychology* 42: 66–75.

Cheng, A. S., L. E. Kruger, and S. E. Daniels. 2003. "'Place' as an Integrating Concept in Natural Resource Politics: Propositions for a Social Science Research Agenda." *Society & Natural Resources* 16(2): 87–104.

Cheng, J. C.-H., and M. C. Monroe. 2012. "Connection to Nature: Children's Affective Attitude toward Nature." *Environment and Behavior* 44(1): 31–49.

Cho, S., and H. Kang. 2017. "Putting Behavior into Context: Exploring the Contours of Social Capital Influences on Environmental Behavior." *Environment and Behavior* 49(3): 283–313.

Chung, H. L., and S. Probert. 2011. "Civic Engagement in Relation to Outcome Expectations among African American Young Adults." *Journal of Applied Developmental Psychology* 32(4): 227–234.

Cialdini, R. B. 2003. "Crafting Normative Messages to Protect the Environment." *Current Directions in Psychological Science* 12(4): 105–109.

——. 2007. "Descriptive Social Norms as Underappreciated Sources of Social Control." *Psychometrika* 72(2): 263.

Cialdini, R. B., S. L. Brown, B. P. Lewis, C. Luce, and S. L. Neuberg. 1997. "Reinterpreting the Empathy-Altruism Relationship: When One into One Equals Oneness." *Journal of Personality and Social Psychology* 73(3): 481–494.

Cialdini, R. B., R. R. Reno, and C. A. Kallgren. 1990. "A Focus Theory of Normative Conduct: Recycling the Concept of Norms to Reduce Littering in Public Places." *Journal of Personality and Social Psychology* 58(6): 1015–1026.

Civic Ecology Lab. 2019a. "Civic Ecology." www.civicecology.org.

——. 2019b. "Global Environmental Education Facebook Group." https://www.facebook.com/groups/GlobalEE/.

Claridge, T. 2004. "Definitions of Social Capital." https://www.socialcapitalresearch.com/literature/definition.html.

Clark, H., and A. Anderson. 2004. *Theories of Change and Logic Models: Telling Them Apart.* Atlanta: American Evaluation Association.

Clark, S., J. E. Petersen, C. M. Frantz, D. Roose, J. Ginn, and D. Rosenberg Daneri. 2017. "Teaching Systems Thinking to 4(th) and 5(th) Graders Using Environmental Dashboard Display Technology." *PLOS ONE* 12(4): e0176322.

Clayton, S. 2003. "Environmental Identity: A Conceptual and Operational Definition." In *Identity and the Natural Environment*, edited by S. Clayton and S. Opotow, 45–65. Cambridge, MA: MIT Press.

——. 2012. "Environment and Identity." In *The Oxford Handbook of Environmental and Conservation Psychology*, edited by S. Clayton, 164–180. Oxford: Oxford University Press.

Coffé, H., and B. Geys. 2007. "Toward an Empirical Characterization of Bridging and Bonding Social Capital." *Nonprofit and Voluntary Sector Quarterly* 36(1): 121–139.

Coleman, J. S. 1988. "Social Capital in the Creation of Human Capital." *American Journal of Sociology* 94, Supplement S95–S120.

Comber, J. 2016. "North Rupununi Wildlife Clubs: Makushi Amerindians' Perceptions of Environmental Education and Positive Youth Development in Guyana." PhD diss., University of Ottawa.

Connell, J. P., and A. C. Kubisch. 1998. "Applying a Theory of Change Approach to the Evaluation of Comprehensive Community Initiatives: Progress, Prospects, and Problems." Washington, DC: Aspen Institute.

Connolly, J. J., E. Svendsen, D. R. Fisher, and L. Campbell. 2014. "Networked Governance and the Management of Ecosystem Services: The Case of Urban Environmental Stewardship in New York City." *Ecosystem Services* 10: 187–194.

Cooper, C. B., J. L. Dickinson, T. Phillips, and R. Bonney. 2007. "Citizen Science as a Tool for Conservation in Residential Ecosystems" *Ecology and Society* 12(2): 11.

Cornelissen, G., M. Pandelaere, L. Warlop, and S. Dewitte. 2008. "Positive Cueing: Promoting Sustainable Consumer Behavior by Cueing Common Environmental Behaviors as Environmental." *International Journal of Research in Marketing* 25(1): 46–55.

Cornell University. 2018. "Habitat Network." http://content.yardmap.org/.

Costantino, T. E., and J. C. Greene. 2003. "Reflections on the Use of Narrative in Evaluation." *American Journal of Evaluation* 24(1): 35–49.

Crocetti, E., P. Jahromi, and W. Meeus. 2012. "Identity and Civic Engagement in Adolescence." *Journal of Adolescence* 35(3): 521–532.

C250. 2004. "Barry Commoner." http://c250.columbia.edu/c250_celebrates/remarkable_columbians/barry_commoner.html.

Cuda, H. S., and E. Glazner. 2015. "The Turtle That Became the Anti-Plastic Straw Poster Child." http://www.plasticpollutioncoalition.org/pft/2015/10/27/the-turtle-that-became-the-anti-plastic-straw-poster-child.

Daily, G. C. 1997. *Nature's Services: Societal Dependence on Natural Ecosystems.* Washington, DC: Island Press.

D'Amato, G. L., and M. Krasny. 2011. "Outdoor Adventure Education: Applying Transformative Learning Theory to Understanding Instrumental Learning and Personal Growth in Environmental Education." *Journal of Environmental Education* 42(4): 237–254.

Danish, J. A. 2014. "Applying an Activity Theory Lens to Designing Instruction for Learning about the Structure, Behavior, and Function of a Honeybee System." *Journal of the Learning Sciences* 23(2): 100–148.

Dart, J., and R. Davies. 2003. "A Dialogical, Story-Based Evaluation Tool: The Most Significant Change Technique." *American Journal of Evaluation* 24(2): 137–155.

Davis, A. C., and M. L. Stroink. 2016. "The Relationship between Systems Thinking and the New Ecological Paradigm." *Systems Research and Behavioral Science* 33(4): 575–586.

Davis, J. L., J. D. Green, and A. Reed. 2009. "Interdependence with the Environment: Commitment, Interconnectedness, and Environmental Behavior." *Journal of Environmental Psychology* 29(2): 173–180.

Deats, R. 2005. *Mahatma Gandhi: Non-violent Liberator; A Biography.* Hyde Park, NY: New City.

DEC (New York Department of Environmental Conservation). n.d. "A Guide to Local Action: Climate Smart Communities Certification." http://www.dec.ny.gov/energy/50845.html.

De Cremer, D., and M. Vugt. 1998. "Collective Identity and Cooperation in a Public Goods Dilemma: A Matter of Trust or Self-Efficacy?" *Current Research in Social Psychology* 3(1): 1–11.

De Groot, J. I. M., and J. Thøgersen. 2013. "Values and Pro-environmental Behavior." In *Environmental Psychology: An Introduction*, edited by L. Steg, A. E. Van den Berg, and J. I. M. de Groot, 141–151. Chichester, UK: John Wiley & Sons.

Delia, J. E. 2013. "Cultivating a Culture of Authentic Care in Urban Environmental Education: Narratives from Youth Interns at East New York Farms!" Master's thesis, Cornell University.

Delia, J. E., and M. E. Krasny. 2018. "Cultivating Positive Youth Development, Critical Consciousness, and Authentic Care in Urban Environmental Education." *Frontiers in Psychology* 8: 2340.

Denchak, M. 2017. "Attention, Online Shoppers." https://www.nrdc.org/stories/attention-online-shoppers.

Derr, V., L. Chawla, and M. Mintzer. 2018. *Placemaking with Children and Youth: Participatory Practices for Planning Sustainable Communities.* New York: New Village.

Design for Public Space. 2015. "Farming Concrete Data Collection Tool." New York: Design for Public Space.

Devine-Wright, P. 2009. "Rethinking NIMBYism: The Role of Place Attachment and Place Identity in Explaining Place-Protective Action." *Journal of Community & Applied Social Psychology* 19(6): 426–441.

Devine-Wright, P., and S. Clayton. 2010. "Introduction to the Special Issue: Place, Identity and Environmental Behaviour." *Journal of Environmental Psychology* 30(3): 267–270.

Devine-Wright, P., and Y. Howes. 2010. "Disruption to Place Attachment and the Protection of Restorative Environments: A Wind Energy Case Study." *Journal of Environmental Psychology* 30(3): 271–280.

de Vreede, C., A. Warner, and R. Pitter. 2014. "Facilitating Youth to Take Sustainability Actions: The Potential of Peer Education." *Journal of Environmental Education* 45(1): 37–56.

Dickinson, J. L. 2009. "The People Paradox: Self-Esteem Striving, Immortality Ideologies, and Human Response to Climate Change." *Ecology & Society* 14(1): 34.

Dickinson, J. L., and R. Bonney, eds. 2012. *Citizen Science: Public Collaboration in Environmental Research.* Ithaca, NY: Cornell University Press.

Dickinson, J. L., R. L. Crain, H. K. Reeve, and J. P. Schuldt. 2013. "Can Evolutionary Design of Social Networks Make It Easier to Be 'Green'?" *Trends in Ecology & Evolution* 28(9): 561–569.

Dieser, O., and F. X. Bogner. 2016. "Young People's Cognitive Achievement as Fostered by Hands-on-Centred Environmental Education." *Environmental Education Research* 22(7): 943–957.

Dietz, M. E., J. C. Clausen, and K. K. Filchak. 2004. "Education and Changes in Residential Nonpoint Source Pollution." *Environmental Management* 34(5): 684–690.

Dietz, T., G. T. Gardner, J. Gilligan, P. C. Stern, and M. P. Vandenbergh. 2009. "Household Actions Can Provide a Behavioral Wedge to Rapidly Reduce US Carbon Emissions." *Proceedings of the National Academy of Sciences* 106(44): 18452–18456.

Dietz, T., E. Ostrom, and P. C. Stern. 2003. "The Struggle to Govern the Commons." *Science* 302(5652): 1907–1912.

Dittmer, L. D., and M. Riemer. 2012. "Fostering Critical Thinking about Climate Change: Applying Community Psychology to an Environmental Education Project with Youth." *Global Journal of Community Psychology Practice* 4(1): 1–12.

Dono, J., J. Webb, and B. Richardson. 2010. "The Relationship between Environmental Activism, Pro-environmental Behaviour and Social Identity." *Journal of Environmental Psychology* 30(2): 178–186.

Dovidio, J., S. Gaertner, and V. Esses. 2007. "Cooperation, Common Identity, and Intergroup Contact." In *Cooperation: The Political Psychology of Effective Human Interaction*, edited by B. Sullivan, M. Snyder, and J. Sullivan, 143–159. New York: Wiley-Blackwell.

Doyle, R., and M. E. Krasny. 2003. "Participatory Rural Appraisal as an Approach to Environmental Education in Urban Community Gardens." *Environmental Education Research* 9(1): 91–115.

Drawdown. n.d. "Solutions." https://www.drawdown.org/solutions.

Duarte, R., J.-J. Escario, and M.-V. Sanagustín. 2017. "The Influence of the Family, the School, and the Group on the Environmental Attitudes of European Students." *Environmental Education Research* 23(1): 23–42.

DuBois, B., and M. E. Krasny. 2016. "Educating with Resilience in Mind: Addressing Climate Change in Post-Sandy New York City." *Journal of Environmental Education* 47(4): 255–270.

DuBois, B., M. E. Krasny, and J. G. Smith. 2017. "Connecting Brawn, Brains, and People: An Exploration of Non-traditional Outcomes of Youth Stewardship Programs." *Environmental Education Research* 24(7): 1–18.

Dunk, T. n.d. "Plogging 101: The Workout That Sheds Kilos and Helps the Environment." Lifestyle Australia.

Dunlap, R. E., and K. D. Van Liere. 1978. "The 'New Environmental Paradigm.'" *Journal of Environmental Education* 9(4): 10–19.

——. 2008. "The 'New Environmental Paradigm.'" *Journal of Environmental Education* 40(1): 19–28.

Dunlap, R. E., K. D. Van Liere, A. G. Mertig, and R. E. Jones. 2000. "Measuring Endorsement of the New Ecological Paradigm: A Revised NEP Scale." *Journal of Social Issues* 56(3): 425–442.

Dutcher, D. D., J. C. Finley, A. E. Luloff, and J. B. Johnson. 2007. "Connectivity with Nature as a Measure of Environmental Values." *Environment and Behavior* 39(4): 474–493.

Dyg, P. M., and K. Wistoft. 2018. "Wellbeing in School Gardens—the Case of the Gardens for Bellies Food and Environmental Education Program." *Environmental Education Research* 24(8): 1–15.

Eames-Sheavly, M., L. J. Brewer, and F. Doherty. 2018. "Cornell Garden-Based Learning." http://gardening.cals.cornell.edu/garden-guidance/.

Earth Force. n.d. "Community Action and Problem-Solving Process: An Overview." http://earthforceresources.org/community-action-and-problem-solving-process-an-overview/.

Eccles, J., and J. A. Gootman. 2002. *Community Programs to Promote Youth Development.* Washington, DC: National Academies Press.

Edelstein, M. 2002. "Contamination: The Invisible Built Environment." In *Handbook of Environmental Psychology*, edited by R. Bechtel and A. Churchman, 559–588. New York: Wiley.

Eilam, E., and T. Trop. 2012. "Environmental Attitudes and Environmental Behaviors: Which Is the Horse and Which Is the Cart?" *Sustainability* 4: 2210–2246.

Eizenberg, E. 2013. *From the Ground Up: Community Gardens in New York City and the Politics of Spatial Transformation*. Farnham, Surrey, UK: Ashgate.

Energy Innovation. 2018. "Energy Policy Simulator." https://www.energypolicy.solutions/.

EnergySave. 2018. "Energy Conservation: 10 Ways to Save Energy." https://www.energysage.com/energy-efficiency/101/ways-to-save-energy/.

Engler, S. 2016. "10 Ways to Reduce Plastic Pollution." https://www.nrdc.org/stories/10-ways-reduce-plastic-pollution.

Ennis, R. H. 1993. "Critical Thinking Assessment." *Theory into Practice* 32(3): 179–186.

Eom, K., H. S. Kim, D. K. Sherman, and K. Ishii. 2016. "Cultural Variability in the Link between Environmental Concern and Support for Environmental Action." *Psychological Science* 27(10): 1331–1339.

Erikson, K. T. 1976. *Everything in Its Path*. New York: Simon & Schuster.

Ernst, J. A., N. Blood, and T. Beery. 2017. "Environmental Action and Student Environmental Leaders: Exploring the Influence of Environmental Attitudes, Locus of Control, and Sense of Personal Responsibility." *Environmental Education Research* 23(2): 149–175.

Ernst, J. A., and M. C. Monroe. 2004. "The Effects of Environment-Based Education on Students' Critical Thinking Skills and Disposition toward Critical Thinking." *Environmental Education Research* 10(4): 507–522.

Ernst, J. A., M. C. Monroe, and B. Simmons. 2009. *Evaluating Your Environmental Education Programs: A Workbook for Practitioners*. Washington, DC: North American Association for Environmental Education.

Ernst, J. A., and S. Theimer. 2011. "Evaluating the Effects of Environmental Education Programming on Connectedness to Nature." *Environmental Education Research* 17(5): 577–598.

Faber, J., A. Schroten, M. Bles, M. Sevenster, A. Markowska, M. Smit, C. Rohde, et al. 2012. *Behavioural Climate Change Mitigation Options and Their Appropriate Inclusion in Quantitative Longer Term Policy Scenarios: Main Report*. Delft: European Comission.

Facione, P. A. 1990. "Critical Thinking: A Statement of Expert Consensus for Purposes of Educational Assessment and Instruction, Executive Summary," Fullerton, CA: American Philosophical Association.

Farnum, J., T. Hall, and L. E. Kruger. 2005. "Sense of Place in Natural Resource Recreation and Tourism: An Evaluation and Assessment of Research Findings." US Forest Service.

Farrow, K., G. Grolleau, and L. Ibanez. 2017. "Social Norms and Pro-environmental Behavior: A Review of the Evidence." *Ecological Economics* 140: 1–13.

Ferguson, K. M. 2006. "Social Capital and Children's Wellbeing: A Critical Synthesis of the International Social Capital Literature." *International Journal of Social Welfare* 15: 2–18.

Fernández-Ballesteros, R., J. Díez-Nicolás, G. V. Caprara, C. Barbaranelli, and A. Bandura. 2002. "Determinants and Structural Relation of Personal Efficacy to Collective Efficacy." *Applied Psychology* 51(1): 107–125.

Festinger, L. 1962. "Cognitive Dissonance." *Scientific American* 207(4): 93–106.
Fielding, K. S., and M. J. Hornsey. 2016. "A Social Identity Analysis of Climate Change and Environmental Attitudes and Behaviors: Insights and Opportunities." *Frontiers in Psychology* 7: 121.
Finger, M. 1994. "From Knowledge to Action: Exploring the Relationships between Environmental Education, Learning, and Behavior." *Journal of Social Issues* 50(3): 141–160.
Fisher, D. R., E. S. Svendsen, and J. J. Connolly. 2015. *Urban Environmental Stewardship and Civic Engagement: How Planting Trees Strengthens the Roots of Democracy.* New York: Routledge.
Flanagan, C., and P. Levine. 2010. "Civic Engagement and the Transition to Adulthood." *Future of Children* 20(1): 159–179.
Folke, C. 2016. "Resilience." *Oxford Research Encyclopedia, Environmental Science*, 1–63. New York: Oxford University Press.
Folke, C., S. Carpenter, T. Elmqvist, L. H. Gunderson, C. S. Holling, B. H. Walker, J. Bengtsson, et al. 2002. "Resilience and Sustainable Development: Building Adaptive Capacity in a World of Transformations." Report from the World Summit on Sustainable Development, Johannesburg.
Folke, C., J. Colding, and F. Berkes. 2003. "Synthesis: Building Resilience and Adaptive Capacity in Social-Ecological Systems." In *Navigating Social-Ecological Systems: Building Resilience for Complexity and Change*," edited by F. Berkes, J. Colding, and C. Folke, 352–365. New York: Cambridge University Press.
Frantz, C. M., and F. S. Mayer. 2014. "The Importance of Connection to Nature in Assessing Environmental Education Programs." *Studies in Educational Evaluation* 41: 85–89.
Fraser, J., and C. B. Brandt. 2013. "The Emotional Life of the Environmental Educator." In *Trading Zones in Environmental Education: Creating Transdisciplinary Dialogue*, edited by M. E. Krasny and J. Dillon, 133–158. New York: Peter Lang.
Fraser, J., S. Clayton, J. Sickler, and A. Taylor. 2009. "Belonging at the Zoo: Retired Volunteers, Conservation Activism and Collective Identity." *Ageing & Society* 29(3): 351–368.
Fraser, J., R. Gupta, and M. E. Krasny. 2015. "Practitioners' Perspectives on the Purpose of Environmental Education." *Environmental Education Research* 21(5): 777–800.
Freire, P. 1970. *Pedagogy of the Oppressed.* New York: Herder and Herder.
——. 1973. *Education for Critical Consciousness.* New York: Continuum.
Fremerey, C., and F. Bogner. 2014. "Learning about Drinking Water: How Important Are the Three Dimensions of Knowledge That Can Change Individual Behavior?" *Education Sciences* 4(4): 213.
Frick, J., F. G. Kaiser, and M. Wilson. 2004. "Environmental Knowledge and Conservation Behavior: Exploring Prevalence and Structure in a Representative Sample." *Personality and Individual Differences* 37(8): 1597–1613.
Fritze, J. G., G. A. Blashki, S. Burke, and J. Wiseman. 2008. "Hope, Despair and Transformation: Climate Change and the Promotion of Mental Health and Wellbeing." *International Journal of Mental Health Systems* 2(1): 13.
Garavito-Bermúdez, D., C. Lundholm, and B. Crona. 2016. "Linking a Conceptual Framework on Systems Thinking with Experiential Knowledge." *Environmental Education Research* 22(1): 89–110.
Gatersleben, B. 2013. "Measuring Environmental Behavior." In *Environmental Psychology: An Introduction*, edited by L. Steg, A. E. Van den Berg, and J. I. M. de Groot, 131–140. Chichester, UK: John Wiley & Sons.

Gatersleben, B., N. Murtagh, and W. Abrahamse. 2014. "Values, Identity and Pro-environmental Behavior." *Contemporary Social Science* 9(4): 374–392.
Geller, E. S. 2002. "The Challenge of Increasing Proenvironment Behavior." In *Handbook of Environmental Psychology*, edited by R. Bechtel and A. Churchman, 525–540. New York: John Wiley & Sons.
Gifford, R. 2014. "Environmental Psychology Matters." *Annual Review of Psychology* 65(1): 541–579.
Gifford, R., and R. Sussman. 2012. "Environmental Attitudes." In *Handbook of Environmental and Conservation Psychology*, edited by S. Clayton, 65–80. Oxford: Oxford University Press.
Ginwright, S., and J. Cammarota. 2002. "New Terrain in Youth Development: The Promise of a Social Justice Approach." *Social Justice* 29(4)(90): 82–95.
Global Digital Citizen Foundation. n.d. "Critical Thinking CheatSheet." https://wabisabilearning.com/resources/?.
Goldstein, N. J., R. B. Cialdini, and V. Griskevicius. 2008. "A Room with a Viewpoint: Using Social Norms to Motivate Environmental Conservation in Hotels." *Journal of Consumer Research* 35(3): 472–482.
Goncalves, M. 2009. "Environmentalists Celebrate Return of Alewife." *Hunts Point Express* (New York), May 28.
Goralnik, L., and M. P. Nelson. 2011. "Framing a Philosophy of Environmental Action: Aldo Leopold, John Muir, and the Importance of Community." *Journal of Environmental Education* 42(3): 181–192.
Gosling, E., and K. J. H. Williams. 2010. "Connectedness to Nature, Place Attachment and Conservation Behaviour: Testing Connectedness Theory among Farmers." *Journal of Environmental Psychology* 30(3): 298–304.
Grant, S., and M. Humphries. 2006. "Critical Evaluation of Appreciative Inquiry: Bridging an Apparent Paradox." *Action Research* 4(4): 401–418.
Gray, S., R. Jordan, A. Crall, G. Newman, C. Hmelo-Silver, J. Huang, W. Novak, et al. 2017. "Combining Participatory Modelling and Citizen Science to Support Volunteer Conservation Action." *Biological Conservation* 208: 76–86.
Green, C., D. Kalvaitis, and A. Worster. 2015. "Recontextualizing Psychosocial Development in Young Children: A Model of Environmental Identity Development." *Environmental Education Research* 22(7): 1–24.
Greene, J. C. 2010. "Serving the Public Good." *Evaluation and Program Planning* 33(2): 197–200.
Grist. 2018. "Ask Umbra." https://grist.org/author/ask-umbra/.
Gruenewald, D. A. 2003. "The Best of Both Worlds: A Critical Pedagogy of Place." *Educational Researcher* 32(4): 3–12.
Gunderson, L. C., E. W. Pierce, and M. E. Krasny. 2018. "Adaptive Management, Adaptive Governance, and Civic Ecology." In Krasny, *Grassroots to Global*, 157–176.
Gunderson, L. H., and C. S. Holling, eds. 2002. *Panarchy: Understanding Transformations in Human and Natural Systems*. Washington DC: Island Press.
Halama, P., and M. Dedova. 2007. "Meaning in Life and Hope as Predictors of Positive Mental Health: Do They Explain Residual Variance Not Predicted by Personality Traits?" *Studia Psychologica* 49(3): 191–200.
Halpenny, E. A. 2010. "Pro-environmental Behaviours and Park Visitors: The Effect of Place Attachment." *Journal of Environmental Psychology* 30(4): 409–421.
Hanifan, L. J. 1916. "The Rural School Community Center." *Annals of the American Academy of Political and Social Science* 67: 130–138.
Hart, R. 1992. *Children's Participation: From Tokenism to Citizenship*. New York: UNICEF.

Haywood, B. K., J. K. Parrish, and J. Dolliver. 2016. "Place-Based and Data-Rich Citizen Science as a Precursor for Conservation Action." *Conservation Biology* 30(3): 476–486.

Health Council of the Netherlands. 2004. "The Influence of Nature on Social, Psychological and Physical Well-Being." The Hague: Health Council of the Netherlands.

Heberlein, T. A. 2012. *Navigating Environmental Attitudes.* Oxford: Oxford University Press.

Heimlich, J. E. 2010. "Environmental Education Evaluation: Reinterpreting Education as a Strategy for Meeting Mission." *Evaluation and Program Planning* 33(2): 180–185.

Helve, H., and J. Bynner. 2007. *Youth and Social Capital.* London: Tufnell.

Hester, R. 1993. "Sacred Spaces and Everyday Life: A Return to Manteo, North Carolina." In *Dwelling, Seeing, and Designing: Toward a Phenomenological Ecology*, edited by D. Seamon, 271–297. Albany: SUNY Press.

Hines, J. M., H. R. Hungerford, and A. N. Tomera. 1986/87. "Analysis and Synthesis of Research on Responsible Environmental Behavior: A Meta-analysis." *Journal of Environmental Education* 18(2): 1–8.

Hinson, L., C. Kapungu, C. Jessee, M. Skinner, M. Bardini, and T. Evans-Whipp. 2016. "Measuring Positive Youth Development Toolkit: A Guide for Implementers of Youth Programs." Washington, DC: YouthPower Learning, Making Cents International.

Hmelo-Silver, C. E., R. Jordan, C. Eberbach, and S. Sinha. 2017. "Systems Learning with a Conceptual Representation: A Quasi-experimental Study." *Instructional Science* 45(1): 53–72.

Hofreiter, T. D., M. C. Monroe, and T. V. Stein. 2007. "Teaching and Evaluating Critical Thinking in an Environmental Context." *Applied Environmental Education & Communication* 6(2): 149–157.

Holland, D., G. Fox, and V. Daro. 2008. "Social Movements and Collective Identity." *Anthropological Quarterly* 81(1): 95–126.

Holland, D., and J. Lave. 2009. "Social Practice Theory and the Historical Production of Persons." *Actio: An International Journal of Human Activity Theory* 2: 1–15.

Holling, C. S. 1973. "Resilience and Stability of Ecological Systems." *Annual Review of Ecology, Evolution, and Systematics* 4: 1–23.

Hollweg, K., J. Taylor, R. W. Bybee, T. Marcinkowski, W. C. McBeth, and P. Zoido. 2011. *Developing a Framework for Assessing Environmental Literacy*, 122. Washington, DC: NAAEE.

Holtan, M. T., S. L. Dieterlen, and W. C. Sullivan. 2014. "Social Life under Cover: Tree Canopy and Social Capital in Baltimore, Maryland." *Environment and Behavior* 47(5): 502–525.

Homburg, A., and A. Stolberg. 2006. "Explaining Pro-environmental Behavior with a Cognitive Theory of Stress." *Journal of Environmental Psychology* 26(1): 1–14.

Howell, A. J., H.-A. Passmore, and K. Buro. 2013. "Meaning in Nature: Meaning in Life as a Mediator of the Relationship between Nature Connectedness and Well-Being." *Journal of Happiness Studies* 14(6): 1681–1696.

Hu, W. 2018. "Could New York City Parks Be Going Plastic Bottle-Free?" *New York Times*, April 20.

Huber, J., W. K. Viscusi, and J. Bell. 2018. "Dynamic Relationships between Social Norms and Pro-environmental Behavior: Evidence from Household Recycling." *Behavioural Public Policy*, February 19, 1–25.

Huckle, J. 1993. "Environmental Education and Sustainability: A View from Critical Theory." In *Environmental Education: A Pathway to Sustainability*, edited by J. Fien, 43–68. Melbourne: Deakin University.

Hungerford, H. R., and T. Volk. 1990. "Changing Learner Behavior through Environmental Education." *Journal of Environmental Education* 21(3): 8–21.

Hurley, D. 2004. "Scientist at Work—Felton Earls; On Crime as Science (a Neighbor at a Time)." *New York Times*, January 6.

Hutchinson, B. 2014. "7 Ways to Reduce Ocean Plastic Pollution Today." https://www.oceanicsociety.org/blog/1720/7-ways-to-reduce-ocean-plastic-pollution-today.

Hwang, Y.-H., S.-I. Kim, and J.-M. Jeng. 2000. "Examining the Causal Relationships among Selected Antecedents of Responsible Environmental Behavior." *Journal of Environmental Education* 31(4): 19–25.

IPCC. 2014. *Climate Change 2014: Impacts, Adaptation, and Vulnerability. Part A: Global and Sectoral Aspects. Contribution of Working Group II to the Fifth Assessment Report of the Intergovernmental Panel on Climate Change*, edited by C. B. Field, V. R. Barros, D. J. Dokken, K. J. Mach, M. D. Mastrandrea, T. E. Bilir, M. Chatterjee, et al. Cambridge: Cambridge University Press.

——. 2018. *Global Warming of 1.5°C: An IPCC Special Report on the Impacts of Global Warming of 1.5°C above Pre-industrial Levels and Related Global Greenhouse Gas Emission Pathways, in the Context of Strengthening the Global Response to the Threat of Climate Change, Sustainable Development, and Efforts to Eradicate Poverty*, edited by V. Masson-Delmotte, P. Zhai, H. O. Pörtner, D. Roberts, J. Skea, P.R. Shukla, A. Pirani, et al. Seoul: IPCC.

IPCC Core Writing Team. 2014. *Climate Change 2014: Synthesis Report; Contribution of Working Groups I, II and III to the Fifth Assessment Report of the Intergovernmental Panel on Climate Change*. Geneva: IPCC.

Jarrett, R. L., P. J. Sullivan, and N. D. Watkins. 2005. "Developing Social Capital through Participation in Organized Youth Programs: Qualitative Insights from Three Programs." *Journal of Community Psychology* 33(1): 41–55.

Jensen, B. B., and K. Schnack. 1997. "The Action Competence Approach in Environmental Education." *Environmental Education Research* 3(2): 163–178.

Jimenez-Aceituno, A., L. Medland, A. Delgado, A. Carballes-Breton, A. Maiques-Diaz, L. D. Munoz, M. Marin-Rodriguez, P. Chamorro-Ortiz, and B. Casado-Cid. 2016. "Social Theatre as a Tool for Environmental Learning Processes: A Case Study from Madrid, Spain." In *Across the Spectrum: Resources for Environmental Educators*, edited by M. Monroe and M. E. Krasny, 281–295. Washington, DC: NAAEE.

Johnson, B., M. Duffin, and M. Murphy. 2012. "Quantifying a Relationship between Place-Based Learning and Environmental Quality." *Environmental Education Research* 18(5): 609–624.

Johnson, L. R., J. S. Johnson-Pynn, and T. M. Pynn. 2007. "Youth Civic Engagement in China: Results from a Program Promoting Environmental Activism." *Journal of Adolescent Research* 22(4): 355–386.

Johnson, R. B., and A. J. Onwuegbuzie. 2004. "Mixed Methods Research: A Research Paradigm Whose Time Has Come." *Educational Researcher* 33(7): 14–26.

Johnson-Pynn, J. S., and L. R. Johnson. 2010. "Exploring Environmental Education for East African Youth: Do Program Contexts Matter?" *Children, Youth and Environments* 20(1): 123–151.

Jorgensen, B. S., and R. C. Stedman. 2001. "Sense of Place as an Attitude: Lakeshore Owners' Attitudes toward Their Properties." *Journal of Environmental Psychology* 21: 233–248.

——. 2006. "A Comparative Analysis of Predictors of Sense of Place Dimensions: Attachment to, Dependence on, and Identification with Lakeshore Properties." *Journal of Environmental Management* 79: 316–327.

JouleBug. 2018. "JouleBug." https://joulebug.com/.

Jugert, P., K. H. Greenaway, M. Barth, R. Büchner, S. Eisentraut, and I. Fritsche. 2016. "Collective Efficacy Increases Pro-environmental Intentions through Increasing Self-Efficacy." *Journal of Environmental Psychology* 48: 12–23.

Kahan, D. M. 2015. "Climate-Science Communication and the Measurement Problem." *Political Psychology* 36: 1–43.

Kaiser, F. G., G. Hübner, and F. X. Bogner. 2005. "Contrasting the Theory of Planned Behavior with the Value-Belief-Norm Model in Explaining Conservation Behavior." *Journal of Applied Social Psychology* 35(10): 2150–2170.

Kaiser, F. G., S. Wölfing, and U. Fuhrer. 1999. "Environmental Attitude and Ecological Behavior." *Journal of Environmental Psychology* 19(1): 1–19.

Kals, E., D. Schumacher, and L. Montada. 1999. "Emotional Affinity toward Nature as a Motivational Basis to Protect Nature." *Environment and Behavior* 31(2): 178–202.

Kassam, K.-A., Z. Golshani, and M. E. Krasny. 2018. "Grassroots Stewardship in Iran: The Rise and Significance of Nature Cleaners." In Krasny, *Grassroots to Global*, 65–84.

Kellert, S. R., and E. O. Wilson, eds. 1993. *The Biophilia Hypothesis*. Washington, DC: Island Press.

Kempton, W., and D. Holland. 2003. "Identity and Sustained Environmental Practice." In *Identity and the Natural Environment*, edited by S. Clayton and S. Opotow, 317–341. Hong Kong: MIT Press.

Kennedy School of Government. 2012. "Saguaro Seminar: Civic Engagement in America." http://www.hks.harvard.edu/saguaro/; http://www.hks.harvard.edu/saguaro/measurement/measurement.htm. Accessed August 27, 2012.

Kinzig, A. P., P. R. Ehrlich, L. J. Alston, K. Arrow, S. Barrett, T. G. Buchman, G. C. Daily, et al. 2013. "Social Norms and Global Environmental Challenges: The Complex Interaction of Behaviors, Values, and Policy." *Bioscience* 63(3): 164–175.

Kitchell, A., E. Hannan, and W. Kempton. 2000. "Identity through Stories: Story Structure and Function in Two Environmental Groups." *Human Organization* 59(1): 96–105.

Kitchell, A., W. Kempton, D. Holland, and D. Tesch. 2000. "Identities and Actions within Environmental Groups." *Human Ecology Review* 7(2): 1–20.

Klyza, C. M., J. Isham, and A. Savage. 2006. "Local Environmental Groups and the Creation of Social Capital: Evidence from Vermont." *Society & Natural Resources* 19(10): 905–919.

Kobori, H., J. L. Dickinson, I. Washitani, R. Sakurai, T. Amano, N. Komatsu, W. Kitamura, et al. 2016. "Citizen Science: A New Approach to Advance Ecology, Education, and Conservation." *Ecological Research* 31(1): 1–19.

Kollmuss, A., and J. Agyeman. 2002. "Mind the Gap: Why Do People Act Environmentally and What Are the Barriers to Pro-environmental Behavior?" *Environmental Education Research* 8(3): 239–260.

Krasny, M. E., ed. 2018. *Grassroots to Global: Broader Impacts of Civic Ecology*. Ithaca, NY: Cornell University Press.

Krasny, M. E. 2019. "Cornell University Experiments with Sustainability Solutions and Science." http://greenubuntu.com/cornell-university-experiments-with-sustaina bility-solutions-and-science/.

Krasny, M. E., C.-H. Chang, M. Hauk, and B. DuBois. 2017. "Climate Change Education." In Russ and Krasny, *Urban Environmental Education Review*, 76–85.

Krasny, M. E., S. R. Crestol, K. G. Tidball, and R. C. Stedman. 2014. "New York City's Oyster Gardeners: Memories, Meanings, and Motivations of Volunteer Environmental Stewards." *Landscape and Urban Planning* 132: 16–25.

Krasny, M. E., E. Danter, Y. Li, and R. Gupta. 2017. "Social Innovation and Environmental Education: Lessons from EPA's National Environmental Training Program." Ithaca, NY: Cornell University Civic Ecology Lab.

Krasny, M. E., and B. DuBois. 2016. "Climate Adaptation Education: Embracing Reality or Abandoning Environmental Values?" *Environmental Education Research*, June 16: 1–12.

Krasny, M. E., L. Kalbacker, R. Stedman, and A. Russ. 2013. "Measuring Social Capital among Youth: Applications in Environmental Education." *Environmental Education Research* 21(1): 1–23.

Krasny, M. E., C. Lundholm, E. Lee, S. Shava, and H. Kobori. 2013. "Urban Landscapes as Learning Arenas for Sustainable Management of Biodiversity and Ecosystem Services." In *Urbanization, Biodiversity and Ecosystem Services: Challenges and Opportunities*, edited by T. Elmqvist, M. Fragkias, J. Goodness, et al., 629–664. New York: Springer.

Krasny, M. E., C. Lundholm, and R. Plummer. 2010. "Resilience, Learning and Environmental Education." *Environmental Education Research* (special issue) 15(5–6): 463–672.

——, eds. 2011. *Resilience in Social-Ecological Systems: The Role of Learning and Education*. New York: Taylor & Francis.

Krasny, M. E., M. Mukute, O. M. Aguilar, M. P. Masilela, and P. Olvitt. 2017. "Community Environmental Education." In Russ and Krasny, *Urban Environmental Education Review*, 124–132.

Krasny, M. E., and W.-M. Roth. 2010. "Environmental Education for Social-Eecological System Resilience: A Perspective from Activity Theory." *Environmental Education Research* 16(5–6): 545–558.

Krasny, M. E., A. Russ, K. G. Tidball, and T. Elmqvist. 2013. "Civic Ecology Practices: Participatory Approaches to Generating and Measuring Ecosystem Services in Cities." *Ecosystem Services* 7: 177–186.

Krasny, M. E., and K. Snyder, eds. 2016. *Civic Ecology: Stories about Love of Life, Love of Place*. Ithaca, NY: Cornell University Civic Ecology Lab.

Krasny, M. E., E. S. Svendsen, and C. Konijnendijk von den Bosch. 2017. "Environmental Governance." In Russ and Krasny, *Urban Environmental Education Review*, 103–111.

Krasny, M. E., and K. G. Tidball. 2009a. "Applying a Resilience Systems Framework to Urban Environmental Education." *Environmental Education Research* 15(4): 465–482.

——. 2009b. "Community Gardens as Contexts for Science, Stewardship, and Civic Action Learning." *Cities and the Environment* 2(1): 8.

——. 2015. *Civic Ecology: Adaptation and Transformation from the Ground Up*. Cambridge, MA: MIT Press.

Krasny, M. E., K. G. Tidball, R. Doyle, and N. Najarian. 2005. "Garden Mosaics Website." https://civicecology.org/outreach/garden-mosaics/.

Krasny, M. E., K. G. Tidball, and N. Sriskandarajah. 2009. "Education and Resilience: Social and Situated Learning among University and Secondary Students." *Ecology and Society* 14(2): 38.

Kudryavtsev, A. 2013. "Urban Environmental Education and Sense of Place." PhD diss., Cornell University.

Kudryavtsev, A., M. E. Krasny, and R. Stedman. 2012. "The Impact of Environmental Education on Sense of Place among Urban Youth." *Ecosphere* 3(4): 29.

Kudryavtsev, A., R. Stedman, and M. E. Krasny. 2011. "Sense of Place in Environmental Education." *Environmental Education Research* 18(2): 229–250.

Kuo, F. E., W. C. Sullivan, R. L. Coley, and L. Brunson. 1998. "Fertile Ground for Community: Inner-City Neighborhood Common Spaces." *American Journal of Community Psychology* 26(6): 823–855.

Kurisu, K. 2015. *Pro-environmental Behaviors*. Tokyo: Springer.

Kyle, V., and L. Kearns. 2018. "The Bitter and the Sweet: Weaving a Tapestry of Migration Stories." In Krasny, *Grassroots to Global*, 41–64.

Lacroix, K. 2018. "Comparing the Relative Mitigation Potential of Individual Pro-environmental Behaviors." *Journal of Cleaner Production* 195: 1398–1407.

Læssøe, J., and M. E. Krasny. 2013. "Participation in Environmental Education: Crossing Boundaries within the Big Tent." In *Trans-disciplinary Perspectives in Environmental Education*, edited by M. E. Krasny and J. Dillon, 11–44. New York: Peter Lang.

Landay, J., and A. Crum. n.d. "Environmental Behavior Change App." https://woods.stanford.edu/environmental-venture-projects/environmental-behavior-change-app.

Larsen, T. B., and J. A. Harrington. 2018. "Developing a Learning Progression for Place." *Journal of Geography* 117(3): 100–118.

Larson, L. R., C. B. Cooper, R. C. Stedman, D. J. Decker, and R. J. Gagnon. 2018. "Place-Based Pathways to Proenvironmental Behavior: Empirical Evidence for a Conservation-Recreation Model." *Society & Natural Resources* 31(8): 871–891.

Larson, L. R., R. C. Stedman, C. B. Cooper, and D. J. Decker. 2015. "Understanding the Multi-dimensional Structure of Pro-environmental Behavior." *Journal of Environmental Psychology* 43: 112–124.

Larson, R. W. 2000. "Toward a Psychology of Positive Youth Development." *American Psychologist* 55(1): 170–183.

Larson, R. W., and R. M. Angus. 2011. "Adolescents' Development of Skills for Agency in Youth Programs: Learning to Think Strategically." *Child Development* 82(1): 277–294.

Lauren, N., K. S. Fielding, L. Smith, and W. R. Louis. 2016. "You Did, So You Can and You Will: Self-Efficacy as a Mediator of Spillover from Easy to More Difficult Pro-environmental Behaviour." *Journal of Environmental Psychology* 48: 191–199.

Lee, F. L. F. 2006. "Collective Efficacy, Support for Democratization, and Political Participation in Hong Kong." *International Journal of Public Opinion Research* 18(3): 297–317.

Leicht, A., J. Heiss, and W. J. Byun, eds. 2018. *Issues and Trends in Education for Sustainable Development*. Paris: UNESCO.

Leopold, A. 1949. *A Sand County Almanac*. New York: Oxford University Press.

Lerner, R. M., and J. V. Lerner. 2011. *The Positive Development of Youth: Report of the Findings from the First Seven Years of the 4-H Study of Positive Youth Development*. Medford, MA: Institute for Applied Research in Youth Development, Tufts University, 22.

Lerner, R. M., V. Lerner, J. B. Almerigi, C. Theokas, E. Phelps, S. Gestsdottir, S. Naudeau, et al. 2005. "Positive Youth Development, Participation in Community Youth Programs, and Community Contributions of Fifth-Grade Adolescents: Findings from the First Wave of the 4-H Study of Positive Youth Development." *Journal of Early Adolescence* 25(1): 17–71.

Lerner, R. M., J. V. Lerner, and J. B. Benson. 2011. "Positive Youth Development: Research and Applications for Promoting Thriving in Adolescence." In *Advances in Child Development and Behavior: Positive Youth Development*, edited by R. M. Lerner, J. V. Lerner, and J. B. Benson, 1–17. Amsterdam: Elsevier.

Levy, B. L. M. 2013. "An Empirical Exploration of Factors Related to Adolescents' Political Efficacy." *Educational Psychology* 33(3): 357–390.

Lewicka, M. 2008. "Place Attachment, Place Identity, and Place Memory: Restoring the Forgotten City Past." *Journal of Environmental Psychology* 28(3): 209–231.

——. 2011. "Place Attachment: How Far Have We Come in the Last 40 Years?" *Journal of Environmental Psychology* 31(3): 207–230.

Lewis-Charp, H., H. C. Yu, S. Soukwmneuth, and L. Lacoe. 2003. "Extending the Reach of Youth Development through Civic Activism: Outcomes of the Youth Leadership for Development Initiative." Tacoma Park, MD: Innovation Center for Community and Youth Development.

Li, C. J., and M. C. Monroe. 2017a. "Development and Validation of the Climate Change Hope Scale for High School Students." *Environment and Behavior* 50(4): 454–479.

——. 2017b. "Exploring the Essential Psychological Factors in Fostering Hope Concerning Climate Change." *Environmental Education Research*, August 21: 1–19.

Lieberman, G. A., and L. L. Hoody. 1998. "Closing the Achievement Gap: Using the Environment as an Integrating Context for Learning." San Diego, CA: State Education and Environment Roundtable.

Liefländer, A. K., G. Fröhlich, F. X. Bogner, and P. W. Schultz. 2013. "Promoting Connectedness with Nature through Environmental Education." *Environmental Education Research* 19(3): 370–384.

Lindberg, E., and G. M. Farkas. 2016. "Much Ado about Nothing? A Multilevel Analysis of the Relationship between Voluntary Associations' Characteristics and Their Members' Generalized Trust." *Journal of Civil Society* 12(1): 33–56.

Litterati. 2017. "Litterati." https://www.litterati.org/.

Loader, B. D., A. Vromen, and M. A. Xenos. 2014. "The Networked Young Citizen: Social Media, Political Participation and Civic Engagement." *Information, Communication & Society* 17(2): 143–150.

Lochner, K., I. Kawachi, and B. P. Kennedy. 1999. "Social Capital: A Guide to Its Measurement." *Health & Place* 5: 259–270.

Lotz-Sisitka, H., A. E. J. Wals, D. Kronlid, and D. McGarry. 2015. "Transformative, Transgressive Social Learning: Rethinking Higher Education Pedagogy in Times of Systemic Global Dysfunction." *Current Opinion in Environmental Sustainability* 16: 73–80.

Louv, R. 2008. *Last Child in the Woods: Saving Our Children from Nature-Deficit Disorder.* New York: Algonquin Books.

Lubell, M. 2002. "Environmental Activism as Collective Action." *Environment and Behavior* 34(4): 431–454.

Lucas, R. E., E. Diener, and E. Suh. 1996. "Discriminant Validity of Well-Being Measures." *Journal of Personality and Social Psychology* 71(3): 616–628.

Lundholm, C., and R. Plummer. 2010. "Resilience and Learning: A Conspectus for Environmental Education." *Environmental Education Research* 16(5): 227–243.

Luthar, S. S., D. Cicchetti, and B. Becker. 2000. "The Construct of Resilience: A Critical Evaluation and Guidelines for Future Work." *Child Development* 71(3): 543–562.

Lyson, T. 2004. *Civic Agriculture: Reconnecting Farm, Food, and Community*. Lebanon, NH: Tufts University Press.

Macias, T., and K. Williams. 2016. "Know Your Neighbors, Save the Planet: Social Capital and the Widening Wedge of Pro-environmental Outcomes." *Environment and Behavior* 48(3): 391–420.

Maio, G. R., J. M. Olson, M. M. Bernard, and M. A. Luke. 2006. "Ideologies, Values, Attitudes, and Behavior." In *Handbook of Social Psychology*, edited by J. Delamater, 283–308. Boston: Springer.

Mak, H. W. 2015. "5 Gamified Environmental Apps for Sustainable Living." http://www.gamification.co/2015/10/14/5-gamified-environmental-apps-for-sustainable-living/.

Manzo, L. C., and D. D. Perkins. 2006. "Finding Common Ground: The Importance of Place Attachment to Community Participation and Planning." *Journal of Planning Literature* 20(4): 335–350.

Masten, A. S., and J. Obradovic. 2008. "Disaster Preparation and Recovery: Lessons from Research on Resilience in Human Development." *Ecology and Society* 13(1): 9.

Masterson, V. A., R. C. Stedman, J. Enqvist, M. Tengö, M. Giusti, D. Wahl, and U. Svedin. 2017. "The Contribution of Sense of Place to Social-Ecological Systems Research: A Review and Research Agenda." *Ecology and Society* 22(1).

Matthews, J., and P. Waterman. 2010. "Sustainable Literacy and Climate Change: Engagement, Partnerships, Projects." In *Universities and Climate Change: Introducing Climate Change to University Programmes*, edited by W. L. Filho, 83–88. Heidelberg: Springer.

Mayer, F. S., and C. M. Frantz. 2004. "The Connectedness to Nature Scale: A Measure of Individuals' Feeling in Community with Nature." *Journal of Environmental Psychology* 24: 503–515.

McKenzie, M. 2000. "How Are Adventure Education Program Outcomes Achieved? A Review of the Literature." *Australian Journal of Outdoor Education* 5(1): 19–28.

——. 2003. "Beyond 'the Outward Bound Process': Rethinking Student Learning." *Journal of Experiential Education* 26(1): 8–23.

McKenzie, M., J. R. Koushik, R. Haluza-Delay, and J. Corwin. 2017. "Environmental Justice." In Russ and Krasny, *Urban Environmental Education Review*, 59–67.

McLellan, R., and S. Steward. 2015. "Measuring Children and Young People's Wellbeing in the School Context." *Cambridge Journal of Education* 45(3): 307–332.

McMillan, D. W., and D. M. Chavis. 1986. "Sense of Community: A Definition and Theory." *Journal of Community Psychology* 14: 6–23.

McPhearson, T., and K. G. Tidball. 2013. "Disturbances in Urban Social-Ecological Systems: Niche Opportunities for Environmental Education." In *Trading Zones in Environmental Education: Creating Transdisciplinary Dialogue*, edited by M. E. Krasny and J. Dillon, 193–230. New York: Peter Lang.

MEA. 2005. *Millennium Ecosystem Assessment: Ecosystems and Human Well-Being: Synthesis*. Washington, DC: Island Press.

Mebratu, D. 1998. "Sustainability and Sustainable Development: Historical and Conceptual Review." *Environmental Impact Assessment Review* 18(6): 493–520.

Meichtry, Y., and L. Harrell. 2002. "An Environmental Education Needs Assessment of K–12 Teachers in Kentucky." *Journal of Environmental Education* 33(3): 21–26.

Meinhold, J. L., and A. J. Malkus. 2005. "Adolescent Environmental Behaviors." *Environment and Behavior* 37(4): 511–532.

Meinzen-Dick, R., M. Di Gregorio, and N. McCarthy. 2004. "Methods for Studying Collective Action in Rural Development." Washington, DC: International Food Policy Research Institute.

Mezirow, J. 2000. *Learning as Transformation*. San Francisco: Jossey-Bass.

Milfont, T. L. 2012. "The Psychology of Environmental Attitudes: Conceptual and Empirical Insights from New Zealand." *Ecopsychology* 4(4): 269–276.

Milfont, T. L., and J. Duckitt. 2010. "The Environmental Attitudes Inventory: A Valid and Reliable Measure to Assess the Structure of Environmental Attitudes." *Journal of Environmental Psychology* 30(1): 80–94.

Miller, E., and L. Buys. 2008. "The Impact of Social Capital on Residential Water-Affecting Behaviors in a Drought-Prone Australian Community." *Society & Natural Resources* 21(3): 244–257.

Moldan, B., S. Janoušková, and T. Hák. 2012. "How to Understand and Measure Environmental Sustainability: Indicators and Targets." *Ecological Indicators* 17: 4–13.

Monarch Watch. 2014. "Monarch Watch." http://www.monarchwatch.org/.

Monroe, M. C. 2010. "Challenges for Environmental Education Evaluation." *Evaluation and Program Planning* 33(2): 194–196.

——. 2012. "The Co-evolution of ESD and EE." *Journal of Education for Sustainable Development* 6(1): 43–47.

Monroe, M. C., and S. B. Allred. 2013. "Building Capacity for Community-Based Natural Resource Management with Environmental Education." In *Trading Zones in Environmental Education: Creating Transdisciplinary Dialogue*, edited by M. E. Krasny and J. Dillon, 45–78. New York: Peter Lang.

Monroe, M. C., and K. C. Nelson. 2004. "The Value of Assessing Public Perceptions: Wildland Fire and Defensible Space." *Applied Environmental Education & Communication* 3(2): 109–117.

Monroe, M. C., D. J. Wojcik, and K. Biedenweg. 2016. "A Variety of Strategies for Environmental Education." In *Across the Spectrum: Resources for Environmental Educators*, edited by M. Monroe and M. E. Krasny, 29–46. Washington, DC: NAAEE.

Mooney, C. 2015. "Meet the 'Clean Cow' Technology That Could Help Fight Climate Change." *Washington Post*, July 31.

Mordock, K., and M. E. Krasny. 2001. "Participatory Action Research: A Theoretical and Practical Framework for Environmental Education." *Journal of Environmental Education* 32(3): 15–20.

Morgan, R. 2018. "McDonald's Tests New Approach to Overlooked Environmental Scourge: The Plastic Straw." CNBC, April 22.

Morgensen, F., and K. Schnack. 2010. "The Action Competence Approach and the 'New' Discourses of Education for Sustainable Development, Competence, and Quality Criteria." *Environmental Education Research* 16(1).

Mortensen, C. R., R. Neel, R. B. Cialdini, C. M. Jaeger, R. P. Jacobson and M. M. Ringel. 2019. "Trending Norms: A Lever for Encouraging Behaviors Performed by the Minority." *Social Psychological and Personality Science* 10(2): 201–210.

Morton, T. A., A. Rabinovich, D. Marshall, and P. Bretschneider. 2011. "The Future That May (or May Not) Come: How Framing Changes Responses to Uncertainty in Climate Change Communications." *Global Environmental Change* 21(1): 103–109.

Mueller-Sims, K. 2016. "Environmental Citizens through Earth Force Evaluation." In *Across the Spectrum: Resources for Environmental Educators*, edited by M. Monroe and M. E. Krasny, 251–260. Washington, DC: NAAEE.

Musick, M. A., J. Wilson, and J. W. B. Bynum. 2000. "Race and Formal Volunteering: The Differential Effects of Class and Religion." *Social Forces* 78(4): 1539–1570.

Musser, L. M., and A. J. Malkus. 1994. "The Children's Attitudes toward the Environment Scale." *Journal of Environmental Education* 25(3): 22.

NAAEE (North American Association for Environmental Education). 1998–2016. *Guidelines for Excellence*. Washington, DC: NAAEE.

——. 2018a. "Environmental Issues Forums." https://naaee.org/our-work/programs/environmental-issues-forums.

——. 2018b. "North American Association for Environmental Education Research Library." https://naaee.org/eepro/research/library.

——. n.d. "About EE and Why It Matters." https://naaee.org/about-us/about-ee-and-why-it-matters.

——. n.d. "Diversity, Equity, and Inclusion." https://naaee.org/eepro/learning/eelearn/what-is-ee/lesson-2/diversity-equity-inclusion.

——. n.d. "Tbilisi Definition Animated." https://naaee.org/eepro/learning/eelearn/what-is-ee/lesson-1/animation.

Nadasdy, P. 2007. "Adaptive Co-management and the Gospel of Resilience." In *Collaboration, Learning, and Multi-level Governance*, edited by D. Armitage, F. Berkes, and N. Doubleday, 208–227. Vancouver: University of British Columbia Press.

National Tree Benefit Calculator. n.d. http://www.treebenefits.com/calculator/.

New York. 2018. "East Side Coastal Resiliency Project." http://www1.nyc.gov/site/escr/index.page.

Nigbur, D., E. Lyons, and D. Uzzell. 2010. "Attitudes, Norms, Identity and Environmental Behaviour: Using an Expanded Theory of Planned Behaviour to Predict Participation in a Kerbside Recycling Programme." *British Journal of Social Psychology* 49(2): 259–284.

Nilsson, A., M. Bergquist, and P. W. Schultz. 2017. "Spillover Effects in Environmental Behaviors, across Time and Context: A Review and Research Agenda." *Environmental Education Research* 23(4): 573–589.

Nisbet, E. K., and J. M. Zelenski. 2011. "Underestimating Nearby Nature: Affective Forecasting Errors Obscure the Happy Path to Sustainability." *Psychological Science* 22(9): 1101–1106.

——. 2013. "The NR-6: A New Brief Measure of Nature Relatedness." *Frontiers in Psychology* 4: 813.

Nisbet, E. K., J. M. Zelenski, and S. A. Murphy. 2009. "The Nature Relatedness Scale: Linking Individuals' Connection with Nature to Environmental Concern and Behavior." *Environment and Behavior* 41(5): 715–740.

Nolan, J. M., P. W. Schultz, R. B. Cialdini, N. J. Goldstein, and V. Griskevicius. 2008. "Normative Social Influence Is Underdetected." *Personality and Social Psychology Bulletin* 34(7): 913–923.

Norris, F. H., S. P. Stevens, B. Pffeferbaum, K. F. Wyche, and R. L. Pfefferbaum. 2008. "Community Resilience as a Metaphor, Theory, Set of Capacities, and Strategy for Disaster Readiness." *American Journal of Community Psychology* 41: 127–150.

Nussbaum, M. C. 2011. *Creating Capabilities: The Human Development Approach*. Cambridge, MA: Harvard University Press.

Ocean Conservancy. 2018. "Fighting for Trash Free Seas." https://oceanconservancy.org/trash-free-seas/plastics-in-the-ocean/.

Offer, S., and B. L. Schneider. 2007. "Children's Role in Generating Social Capital." *Social Forces* 85(3): 1125–1142.

Ojala, M. 2012a. "Hope and Climate Change: The Importance of Hope for Environmental Engagement among Young People." *Environmental Education Research* 18(5): 625–642.

——. 2012b. "How Do Children Cope with Global Climate Change? Coping Strategies, Engagement, and Well-Being." *Journal of Environmental Psychology* 32(3): 225–233.

——. 2013. "Coping with Climate Change among Adolescents: Implications for Subjective Well-Being and Environmental Engagement." *Sustainability* 5(5): 2191.

——. 2015. "Hope in the Face of Climate Change: Associations with Environmental Engagement and Student Perceptions of Teachers' Emotion Communication Style and Future Orientation." *Journal of Environmental Education* 46(3): 133–148.

100 Resilient Cities. 2018. "100 Resilient Cities." http://www.100resilientcities.org/.

Okvat, H., and A. Zautra. 2014. "Sowing Seeds of Resilience: Community Gardening in a Post-disaster Context." In *Greening in the Red Zone*, edited by K. G. Tidball and M. E. Krasny, 73–90. New York: Springer.

Orr, A. 2012. "Trash Dance." http://trashdancemovie.com/.

Ostrom, E. 1990. *Governing the Commons*. New York: Cambridge University Press.
——. 2010a. "Analyzing Collective Action." *International Association of Agricultural Economists* 41(s1): 155–166.
——. 2010b. "Polycentric Systems for Coping with Collective Action and Global Environmental Change." *Global Environmental Change* 20: 550–557.
O'Sullivan, E. 2002. "The Project and Vision of Transformative Education: Integral Transformative Learning." In *Expanding the Boundaries of Transformative Learning*, edited by E. O'Sullivan, A. Morrell, and M. A. O'Connor, 1–12. New York: Palgrave.
Otto, S., and P. Pensini. 2017. "Nature-Based Environmental Education of Children: Environmental Knowledge and Connectedness to Nature, Together, Are Related to Ecological Behaviour." *Global Environmental Change* 47: 88–94.
OWC. 2017. "Oldman Watershed Council." http://oldmanwatershed.ca/recreation-overview.
Palmer, B. 2016. "Should You Buy Carbon Offsets? A Practical and Philosophical Guide to Neutralizing Your Carbon Footprint." https://www.nrdc.org/stories/should-you-buy-carbon-offsets.
Patton, M. Q. 2002. *Qualitative Research & Evaluation Methods*. Thousand Oaks, CA: Sage.
Payton, M. A., D. C. Fulton, and D. H. Anderson. 2005. "Influence of Place Attachment and Trust on Civic Action: A Study at Sherburne National Wildlife Refuge." *Society & Natural Resources* 18(6): 511–528.
Peters, K., B. Elands, and A. Buijs. 2010. "Social Interactions in Urban Parks: Stimulating Social Cohesion?" *Urban Forestry & Urban Greening* 9(2): 93–100.
Petersen, J. E., C. M. Frantz, M. R. Shammin, T. M. Yanisch, E. Tincknell, and N. Myers. 2015. "Electricity and Water Conservation on College and University Campuses in Response to National Competitions among Dormitories: Quantifying Relationships between Behavior, Conservation Strategies and Psychological Metrics." *PLOS ONE* 10(12): e0144070.
Petersen, J. E., D. Rosenberg Daneri, C. Frantz, and M. R. Shammin. 2017. "Environmental Dashboards: Fostering Pro-environmental and Pro-community Thought and Action through Feedback." In *Handbook of Theory and Practice of Sustainable Development in Higher Education*, vol. 3, edited by W. Leal Filho, M. Mifsud, C. Shiel, and R. Pretorius, 149–168. Cham, Switzerland: Springer International.
Plummer, R. 2009. "The Adaptive Co-management Process: An Initial Synthesis of Representative Models and Influential Variables." *Ecology and Society* 14(2): 24.
Plummer, R., and J. FitzGibbon. 2007. "Connecting Adaptive Co-management, Social Learning, and Social Capital through Theory and Practice." In *Adaptive Co-Management: Collaboration, Learning, and Multi-level Governance*, edited by D. Armitage, F. Berkes, and N. Doubleday, 38–61. Vancouver: University of British Columbia Press.
Polletta, F., and J. M. Jasper. 2001. "Collective Identity and Social Movements." *Annual Review of Sociology* 27: 283–305.
Portes, A. 1998. "Social Capital: Its Origins and Applications in Modern Sociology." *Annual Review of Sociology* 24: 1–24.
Powell, R. B., M. J. Stern, B. D. Krohn, and N. M. Ardoin. 2011. "Development and Validation of Scales to Measure Environmental Responsibility, Character Development, and Attitudes toward School." *Environmental Education Research* 17(1): 91–111.
Pretty, G. H., H. M. Chipuer, and P. Bramston. 2003. "Sense of Place amongst Adolescents and Adults in Two Rural Australian Towns: The Discriminating Features

of Place Attachment, Sense of Community and Place Dependence in Relation to Place Identity." *Journal of Environmental Psychology* 23(3): 273–287.

Pretty, J. 2003. "Social Capital and the Collective Management of Resources." *Science* 302(5652): 1912–1914.

Price, A., B. Simmons, and M. E. Krasny. 2014. *Principles of Excellence in Community Environmental Education*. Washington, DC: NAAEE.

Primack, R. B., H. Kobori, and S. Mori. 2000. "Dragonfly Pond Restoration Promotes Conservation Awareness in Japan." *Conservation Biology* 14(5): 1153–1154.

Project FeederWatch. 2010. "It's Feeder Season—Time to FeederWatch!" *All About Birds*. http://www.birds.cornell.edu/page.aspx?pid=1964.

Proshansky, H. M., A. K. Fabian, and R. Kaminoff. 1983. "Place-Identity: Physical World Socialization of the Self." *Journal of Environmental Psychology* 3(1): 57–83.

Putnam, R. B. 1995. "Bowling Alone: America's Declining Social Capital." *Journal of Democracy* 6(1): 65–78.

Rakow, D. 2018. "Nature Rx @Cornell." https://naturerx.cornell.edu/default.

Randle, J. M. 2014. "The Systems Thinking Paradigm and Higher-Order Cognitive Processes." Master's thesis, Lakehead University.

Randler, C., and F. X. Bogner. 2009. "Efficacy of Two Different Instructional Methods Involving Complex Ecological Content." *International Journal of Science and Mathematics Education* 7(2): 315–337.

Rebuild by Design. 2015. "BIG U." http://www.rebuildbydesign.org/project/big-team-final-proposal/.

Reed, J. 2007. *Appreciative Inquiry: Research for Change*. Thousand Oaks, CA: Sage.

Reese, A. 2018. "As Countries Crank Up the AC, Emissions of Potent Greenhouse Gases Are Likely to Skyrocket." *Science*, March 8. doi:10.1126/science.aat5331.

Reese, G., and E. Junge. 2017. "Keep on Rockin' in a (Plastic-)Free World: Collective Efficacy and Pro-Environmental Intentions as a Function of Task Difficulty." *Sustainability* 9(2): 200.

Reid, A., B. B. Jensen, and J. Nikel. 2008. *Participation and Learning: Perspectives on Education and the Environment, Health and Sustainability*. New York: Springer.

Reisman, J., and A. Gienapp. 2004. "Theory of Change: A Practical Tool for Action, Results and Learning." Seattle: Organizational Research Services.

Resilience Alliance. 2009. "Resilience Alliance." http://www.resalliance.org/1.php. Accessed September 22, 2009.

——. 2010. *Assessing Resilience in Social-Ecological Systems: Workbook for Practitioners*. Version 2.0, 54. Stockholm: Resilience Alliance.

Reynolds, K., and N. Cohen. 2016. *Beyond the Kale: Urban Agriculture and Social Justice Activism in New York City*. Athens: University of Georgia Press.

Rickard, L. N., and R. C. Stedman. 2015. "From Ranger Talks to Radio Stations: The Role of Communication in Sense of Place." *Journal of Leisure Research* 47(1): 15–33.

Riemer, M., J. Lynes, and G. Hickman. 2014. "A Model for Developing and Assessing Youth-Based Environmental Engagement Programmes." *Environmental Education Research* 20(4): 552–574.

Riemer, M., C. Voorhees, L. Dittmer, S. Alisat, N. Alam, R. Sayal, S. H. Bidisha, et al. 2016. "The Youth Leading Environmental Change Project: A Mixed-Method Longitudinal Study across Six Countries." *Ecopsychology* 8(3): 174–187.

Rioux, L. 2011. "Promoting Pro-environmental Behaviour: Collection of Used Batteries by Secondary School Pupils." *Environmental Education Research* 17(3): 353–373.

Rivlin, L. G. 1982. "Group Membership and Place Meanings in an Urban Neighborhood." *Journal of Social Issues* 38(3): 75–93.

Robinson, T. Y. 2005. "A Study of the Effectiveness of Environmental Education Curricula in Promoting Middle School Students' Critical Thinking Skills." PhD diss., Southern Illinois University.

Roczen, N., F. G. Kaiser, F. X. Bogner, and M. Wilson. 2014. "A Competence Model for Environmental Education." *Environment and Behavior* 46(8): 972–992.

Roper Center for Public Opinion Research. 2000. "Social Capital Benchmark Survey." Cambridge, MA: Saguaro Seminar at John F. Kennedy School of Government, Harvard University.

Ross, L. E. E., K. Arrow, R. Cialdini, N. Diamond-Smith, J. Diamond, J. Dunne, M. Feldman, et al. 2016. "The Climate Change Challenge and Barriers to the Exercise of Foresight Intelligence." *BioScience* 66(5): 363–370.

Rossteutscher, S. 2008. "Social Capital and Civic Engagement: A Comparative Perspective." In *The Handbook of Social Capital*, edited by D. Castiglione, J. W. van Deth, and G. Wolleb, 208–240. Oxford: Oxford University Press.

Roth, J. L., and J. Brooks-Gunn. 2003. "What Is a Youth Development Program? Identification of Defining Principles." In *Handbook of Applied Developmental Science: Promoting Positive Child, Adolescent, and Family Development through Research, Policies, and Programs*, edited by F. Jacobs, D. Wertlieb, and R. M. Lerner, 2: 197–223. Thousand Oaks, CA: Sage.

RTB. 2012. "Rocking the Boat." http://www.rockingtheboat.org/. Accessed August 27, 2012.

Ruepert, A. M., K. Keizer, and L. Steg. 2017. "The Relationship between Corporate Environmental Responsibility, Employees' Biospheric Values and Pro-environmental Behaviour at Work." *Journal of Environmental Psychology* 54: 65–78.

Russ, A., and M. E. Krasny. 2017. *Urban Environmental Education Review*. Ithaca, NY: Cornell University Press.

Russ, A., Y. Li, and M. E. Krasny. 2017. "Urban Environmental Education: Online Course Report." Ithaca, NY: Cornell University Civic Ecology Lab.

Russ, A., S. P. Peters, M. E. Krasny, and R. C. Stedman. 2015. "Development of Ecological Place Meaning in New York City." *Journal of Environmental Education* 46(2): 73–93.

Saldivar, L., and M. E. Krasny. 2004. "The Role of NYC Latino Community Gardens in Community Development, Open Space, and Civic Agriculture." *Agriculture and Human Values* 21: 399–412.

Salusky, I., R. W. Larson, A. Griffith, J. Wu, M. Raffaelli, N. Sugimura, and M. Guzman. 2014. "How Adolescents Develop Responsibility: What Can Be Learned from Youth Programs." *Journal of Research on Adolescence* 24(3): 417–430.

Sampson, R. J., S. W. Raudenbush, and F. Earls. 1997. "Neighborhoods and Violent Crime: A Multilevel Study of Collective Efficacy." *Science* 277(5328): 918–924.

Samsuddin, J., H. Hamisah, and L. C. Ching. 2016. "Digital Engagement, Political and Civic Participation: Mobilizing Youth in Marginalized Communities." *Journal of Business and Social Review in Emerging Economies* 2(1): 14–21.

Sandell, K., and J. Öhman. 2010. "Educational Potentials of Encounters with Nature: Reflections from a Swedish Outdoor Perspective." *Environmental Education Research* 16(1): 113–132.

Saunders, C. 2008. "Double-Edged Swords? Collective Identity and Solidarity in the Environment Movement." *British Journal of Sociology* 59(2): 227–253.

Saylan, C., and D. T. Blumstein. 2011. *The Failure of Environmental Education (and How We Can Fix It*. Berkeley: University of California Press.

Scannell, L., and R. Gifford. 2010. "The Relations between Natural and Civic Place Attachment and Pro-environmental Behavior." *Journal of Environmental Psychology* 30(3): 289–297.

——. 2017. "The Experienced Psychological Benefits of Place Attachment." *Journal of Environmental Psychology* 51: 256–269.

Schönfelder, M. L., and F. X. Bogner. 2017. "Two Ways of Acquiring Environmental Knowledge: By Encountering Living Animals at a Beehive and by Observing Bees via Digital Tools." *International Journal of Science Education* 39(6): 723–741.

Schuitema, G., and C. J. Bergstad. 2013. "Acceptablity of Environmental Policies." In *Environmental Psychology: An Introduction*, edited by L. Steg, A. E. Van den Berg, and J. I. M. de Groot, 255–266. Chichester, UK: John Wiley & Sons.

Schultz, P. W. 2001. "The Structure of Environmental Concern: Concern for Self, Other People, and the Biosphere." *Journal of Environmental Psychology* 21: 327–339.

Schultz, P. W., J. M. Nolan, R. B. Cialdini, N. J. Goldstein, and V. Griskevicius. 2007. "The Constructive, Destructive, and Reconstructive Power of Social Norms." *Psychological Science* 18(5): 429–434.

Schultz, P. W., and S. Oskamp. 1996. "Effort as a Moderator of the Attitude-Behavior Relationship: General Environmental Concern and Recycling." *Social Psychology Quarterly* 59(4): 375–383.

Schultz, P. W., C. Shriver, J. J. Tabanico, and A. M. Khazian. 2004. "Implicit Connections with Nature." *Journal of Environmental Psychology* 24(1): 31–42.

Schultz, P. W., and J. Tabanico. 2007. "Self, Identity, and the Natural Environment: Exploring Implicit Connections with Nature." *Journal of Applied Social Psychology* 37(6): 1219–1247.

Schulz, W. 2005. "Political Efficacy and Expected Political Participation among Lower and Upper Secondary Students: A Comparative Analysis with Data from the IEA Civic Education Study." Budapest: ECPR General Conference.

Schulz, W., and H. Sibberns. 2004. *IEA Civic Education Study Technical Report*, 283. Amsterdam: International Association for the Evaluation of Educational Achievement.

Schusler, T. M. 2014. "Environmental Action and Positive Youth Development." In *Across the Spectrum: Resources for Environmental Educators*, edited by M. Monroe and M. E. Krasny, 107–130. Washington, DC: NAAEE.

Schusler, T. M., and M. E. Krasny. 2008. "Youth Participation in Local Environmental Action: Developing Political and Scientific Literacy." In *Participation and Learning: Perspectives on Education and the Environment, Health and Sustainability*, edited by A. Reid, B. B. Jensen, J. Nikel, and V. Simovska, 268–284. New York: Springer.

——. 2010. "Environmental Action as Context for Youth Development." *Journal of Environmental Education* 41(4): 208–223.

——. 2014. "Science and Democracy in Youth Environmental Action—Learning 'Good' Thinking." In *EcoJustice, Citizen Science and Youth Activism: Situated Tensions for Science Education*, edited by M. Mueller and D. Tippins, 363–384. Basel: Springer.

Schusler, T. M., M. E. Krasny, and D. J. Decker. 2017. "The Autonomy-Authority Duality of Shared Decision-Making in Youth Environmental Action." *Environmental Education Research* 23(4): 533–552.

Schusler, T. M., M. E. Krasny, S. J. Peters, and D. J. Decker. 2009. "Developing Citizens and Communities through Youth Environmental Action." *Environmental Education Research* 15(1): 111–127.

Schwartz, S. H. 1977. "Normative Influences on Altruism." In *Advances in Experimental Social Psychology*, edited by L. Berkowitz, 10: 221–279. Cambridge, MA: Academic Press.

——. 1992. "Universals in the Content and Structure of Values: Theoretical Advances and Empirical Tests in 20 Countries." In *Advances in Experimental Social Psychology*, edited by M. P. Zanna, 25: 1–65. Cambridge, MA: Academic Press.

SEER. 1985. "California Student Assessment Project: Phase Two." Sacramento.

Sellmann, D., and F. X. Bogner. 2013. "Climate Change Education: Quantitatively Assessing the Impact of a Botanical Garden as an Informal Learning Environment." *Environmental Education Research* 19(4): 415–429.

Serriere, S. C. 2014. "The Role of the Elementary Teacher in Fostering Civic Efficacy." *Social Studies* 105(1): 45–56.

Shah, D. V., J. Cho, J. William P. Eveland, and N. Kwak. 2005. "Information and Expression in a Digital Age: Modeling Internet Effects on Civic Participation." *Communication Research* 32(5): 531–565.

Shihui, F., H. Liaquat, and P. Douglas. 2018. "Harnessing Informal Education for Community Resilience." *Disaster Prevention and Management: An International Journal* 27(1): 43–59.

Shirk, J., H. Ballard, C. Wilderman, T. Phillips, A. Wiggins, R. Jordan, E. McCallie, et al. 2012. "Public Participation in Scientific Research: A Framework for Deliberate Design." *Ecology and Society* 17(2): 29.

Short, P. C. 2007. "Use of the Oslo-Potsdam Solution to Test the Effect of an Environmental Education Model on Tangible Measures of Environmental Protection." PhD diss., Southern Illinois University.

——. 2010. "Responsible Environmental Action: Its Role and Status in Environmental Education and Environmental Quality." *Journal of Environmental Education* 41(1): 7–21.

Sierra Club. 2018. "John Muir." https://vault.sierraclub.org/john_muir_exhibit/writings/misquotes.aspx.

Silva, P., and R. L. Ramirez. 2018. "Making Knowledge in Civic Ecology Practices: A Community Garden Case Study." In Krasny, *Grassroots to Global*, 124–140.

Simmons, B. 2004. *Designing Evaluation for Education Projects*. Washington, DC: NOAA.

Sirianni, C., and L. A. Friedland. 2005. *The Civic Renewal Movement: Community Building and Democracy in the United States*. Dayton, OH: Charles F. Kettering Foundation.

Smith, G. A., and D. Sobel. 2010. *Place- and Community-Based Education in Schools*. Exeter, UK: Taylor & Francis.

Smith, J. G., B. DuBois, and M. E. Krasny. 2015. "Framing Resilience through Social Learning: Impacts of Environmental Stewardship on Youth in Post-disturbance Communities." *Sustainability Science* 1(3): 441–453.

Snyder, C. R., K. L. Rand, and D. R. Sigmon. 2018. "Hope Theory: A Member of the Positive Psychology Family." In *The Oxford Handbook of Hope*, edited by M. W. Gallagher and S. J. Lopez, 1–36. Oxford: Oxford University Press.

Sobel, D. 2004. *Place-Based Education: Connecting Classrooms and Communities*. Great Barrington, MA: Orion Society.

Sparkman, G., and G. M. Walton. 2017. "Dynamic Norms Promote Sustainable Behavior, Even If It Is Counternormative." *Psychological Science* 28(11): 1663–1674.

Sriskandarajah, N., R. Bawden, C. Blackmore, K. G. Tidball, and A. E. J. Wals. 2010. "Resilience in Learning Systems: Case Studies in University Education." *Environmental Education Research* 16(5–6): 559–573.

Stapleton, S. 2015. "Environmental Identity Development through Social Interactions, Action, and Recognition." *Journal of Environmental Education* 46(2): 94–113.

Stapp, W. B., A. E. J. Wals, and S. L. Stankorb. 1996. *Environmental Education for Empowerment: Action Research and Community Problem Solving*. Dubuque, IA: Kendall Hunt.

Stedman, R. C. 2002. "Toward a Social Psychology of Place: Predicting Behavior from Place-Based Cognitions, Attitude, and Identity." *Environment and Behavior* 34: 561–581.

Stefaniak, A., M. Bilewicz, and M. Lewicka. 2017. "The Merits of Teaching Local History: Increased Place Attachment Enhances Civic Engagement and Social Trust." *Journal of Environmental Psychology* 51: 217–225.

Steg, L. 2016. "Values, Norms, and Intrinsic Motivation to Act Proenvironmentally." *Annual Review of Environment and Resources* 41(1): 277–292.

Steg, L., J. W. Bolderdijk, K. Keizer, and G. Perlaviciute. 2014. "An Integrated Framework for Encouraging Pro-environmental Behaviour: The Role of Values, Situational Factors and Goals." *Journal of Environmental Psychology* 38: 104–115.

Steg, L., and J. I. M. de Groot. 2010. "Explaining Prosocial Intentions: Testing Causal Relationships in the Norm Activation Model." *British Journal of Social Psychology* 49(4): 725–743.

——. 2012. "Environmental Values." In *The Oxford Handbook of Environmental and Conservation Psychology*, edited by S. Clayton. Oxford: Oxford University Press.

Steg, L., J. I. M. de Groot, L. Dreijerink, W. Abrahamse, and F. Siero. 2011. "General Antecedents of Personal Norms, Policy Acceptability, and Intentions: The Role of Values, Worldviews, and Environmental Concern." *Society & Natural Resources* 24(4): 349–367.

Steg, L., and A. Nordlund. 2013. "Models to Explain Environmental Behavior." In *Environmental Psychology: An Introduction*, edited by L. Steg, A. E. Van den Berg, and J. I. M. de Groot, 185–196. Chichester, UK: John Wiley & Sons.

Steg, L., G. Perlaviciute, E. van der Werff, and J. Lurvink. 2014. "The Significance of Hedonic Values for Environmentally Relevant Attitudes, Preferences, and Actions." *Environment and Behavior* 46(2): 163–192.

Steg, L., and C. Vlek. 2009. "Encouraging Pro-environmental Behaviour: An Integrative Review and Research Agenda." *Journal of Environmental Psychology* 29: 309–317.

Stephens, A. K. 2015. "Developing Environmental Action Competence in High School Students: Examining the California Partnership Academy Model." PhD diss., University of California–Davis.

Sterling, S. 2009. "Towards Sustainable Education." *Environmental Scientist* 18(1): 19–21.

——. 2010. "Learning for Resilience, or the Resilient Learner? Towards a Necessary Reconciliation in a Paradigm of Sustainable Education." *Environmental Education Research* 15(5–6): 511–528.

Stern, M., and A. Hellquist. 2017. "Trust and Collaborative Governance." In Russ and Krasny, *Urban Environmental Education Review*, 94–102.

Stern, M. J., R. B. Powell, and N. M. Ardoin. 2010. "Evaluating a Constructivist and Culturally Responsive Approach to Environmental Education for Diverse Audiences." *Journal of Environmental Education* 42(2): 109–122.

Stern, M. J., R. B. Powell, and D. Hill. 2014. "Environmental Education Program Evaluation in the New Millennium: What Do We Measure and What Have We Learned?" *Environmental Education Research* 20(5): 581–611.

Stern, P. C. 2000a. "Psychology and the Science of Human-Environment Interactions." *American Psychologist* 55(5): 523–530.

——. 2000b. "Toward a Coherent Theory of Environmentally Significant Behavior." *Journal of Social Issues* 56(3): 407–424.

Stern, P. C., and T. Dietz. 1994. "The Value Basis of Environmental Concern." *Journal of Social Issues* 50(3): 65–84.

Stern, P. C., T. Dietz, T. Abel, G. A. Guagnano, and L. Kalof. 1999. "A Value-Belief-Norm Theory of Support for Social Movements: The Case of Environmentalism." *Research in Human Ecology* 6(2): 81–97.

Stern, P. C., T. Dietz, and G. A. Guagnano. 1995. "The New Ecological Paradigm in Social-Psychological Context." *Environment and Behavior* 27(6): 723–743.

Stern, P. C., L. Kalof, T. Dietz, and G. A. Guagnano. 1995. "Values, Beliefs, and Proenvironmental Action: Attitude Formation toward Emergent Attitude Objects." *Journal of Applied Social Psychology* 25(18): 1611–1636.

Stevenson, K., and N. Peterson. 2016. "Motivating Action through Fostering Climate Change Hope and Concern and Avoiding Despair among Adolescents." *Sustainability* 8(1): 6.

Strecher, V. J., B. M. DeVellis, M. H. Becker, and I. M. Rosenstock. 1986. "The Role of Self-Efficacy in Achieving Health Behavior Change." *Health Education Quarterly* 13(1): 73–92.

Sukarieh, M., and S. Tannock. 2011. "The Positivity Imperative: A Critical Look at the 'New' Youth Development Movement." *Journal of Youth Studies* 14(6): 675–691.

Sullivan, B. L., T. Phillips, A. A. Dayer, C. L. Wood, A. Farnsworth, M. J. Iliff, I. J. Davies, et al. 2017. "Using Open Access Observational Data for Conservation Action: A Case Study for Birds." *Biological Conservation* 208: 5–14.

Sullivan, W., F. Kuo, and S. F. DePooter. 2004. "The Fruit of Urban Nature: Vital Neighborhood Spaces." *Environment and Behavior* 36(5): 678–700.

Sundeen, R. A. 1992. "Differences in Personal Goals and Attitudes among Volunteers." *Nonprofit and Voluntary Sector Quarterly* 21(3): 271–291.

Sundeen, R. A., C. Garcia, and S. A. Raskoff. 2009. "Ethnicity, Acculturation, and Volunteering to Organizations." *Nonprofit and Voluntary Sector Quarterly* 38(6): 929–955.

Sunstein, C. R., and L. A. Reisch. 2014. "Automatically Green: Behavioral Economics and Environmental Protection." *Harvard Environmental Law Review* 38(1): 127–158.

Sutton, S. G., and E. Gyuris. 2015. "Optimizing the Environmental Attitudes Inventory: Establishing a Baseline of Change in Students' Attitudes." *International Journal of Sustainability in Higher Education* 16(1): 16–33.

Svendsen, E. S., and L. Campbell. 2008. "Urban Ecological Stewardship: Understanding the Structure, Function and Network of Community-Based Land Management." *Cities and the Environment* 1(1): 1–32.

Syvertsen, A. K., L. Wray-Lake, and A. Metzger. 2015. *Youth Civic and Character Measures Toolkit.* Minneapolis: Search Institute.

Tajfel, H., and J. C. Turner. 1986. "The Social Identity Theory of Intergroup Behavior." In *Psychology of Intergroup Relations*, edited by S. Worchel and L. W. Austin, 7–24. Chicago: Nelson Hall.

Tal, T. 2005. "Implementing Multiple Assessment Modes in an Interdisciplinary Environmental Education Course." *Environmental Education Research* 11(5): 575–601.

Tam, K.-P. 2013. "Concepts and Measures Related to Connection to Nature: Similarities and Differences." *Journal of Environmental Psychology* 34: 64–78.

Taplin, D. H., and H. Clark. 2012. *Theory of Change Basics: A Primer on Theory of Change.* New York: ActKnowledge.

Taplin, D. H., H. Clark, E. Collins, and D. C. Colby. 2013. "Theory of Change: Technical Papers." April. New York: ActKnowledge.

Taylor, D. 2015. "Gender and Racial Diversity in Environmental Organizations: Uneven Accomplishments and Cause for Concern." *Environmental Justice* 8(5): 165–180.

——. 2016. *The Rise of the American Conservation Movement: Power, Privilege and Environmental Protection.* Durham, NC: Duke University Press.

TeachThought. 2018. "48 Critical Thinking Questions for Any Content Area." https://www.teachthought.com/critical-thinking/48-critical-thinking-questions-any-content-area/.

Terry, D. J., M. A. Hogg, and K. M. White. 1999. "The Theory of Planned Behaviour: Self-Identity, Social Identity and Group Norms." *British Journal of Social Psychology* 38: 225–244.

Thaler, R. H., and C. R. Sunstein. 2008. *Nudge: Improving Decisions about Health, Wealth, and Happiness*. New Haven, CT: Yale University Press.
Thibodeau, P. H., C. M. Frantz, and M. L. Stroink. 2016. "Situating a Measure of Systems Thinking in a Landscape of Psychological Constructs." *Systems Research and Behavioral Science* 33(6): 753–769.
Thøgersen, J. 2008. "Social Norms and Cooperation in Real-Life Social Dilemmas." *Journal of Economic Psychology* 29(4): 458–472.
Thøgersen, J., and F. Ölander. 2006. "The Dynamic Interaction of Personal Norms and Environment-Friendly Buying Behavior: A Panel Study." *Journal of Applied Social Psychology* 36(7): 1758–1780.
Thomashow, M. 1995. *Ecological Identity: Becoming a Reflective Environmentalist*. Cambridge, MA: MIT Press.
Thornton, T., and J. Leahy. 2012. "Changes in Social Capital and Networks: A Study of Community-Based Environmental Management through a School-Centered Research Program." *Journal of Science Education and Technology* 21(1): 167–182.
Tidball, K. G., and M. E. Krasny. 2009. "An Ecology of Environmental Education." World Environmental Education Conference, Montreal.
——. 2010. "Urban Environmental Education from a Social-Ecological Perspective: Conceptual Framework." *Cities and the Environment* 3(1): 11.
Tidball, K. G., S. Metcalf, M. Bain, and T. Elmqvist. 2017. "Community-Led Reforestation: Cultivating the Potential of Virtuous Cycles to Confer Resilience in Disaster Disrupted Social-Ecological Systems." *Sustainability Science* 13(3): 797–813.
Torney-Purta, J., R. Lehmann, H. Oswald, and W. Schulz. 2001. *Citizenship and Education in Twenty-Eight Countries*. Amsterdam: International Association for the Evaluation of Educational Achievement.
Truelove, H. B., and A. J. Gillis. 2018. "Perception of Pro-environmental Behavior." *Global Environmental Change* 49: 175–185.
Tuan, Y.-F. 1974. *Topophilia*. Englewood Cliffs, NJ: Prentice-Hall.
TUI. 2010. "The Ugly Indian." http://www.theuglyindian.com/.
UNESCO. 1978. *Intergovernmental Conference on Environmental Education: Final Report*, 96. Paris: UNESCO.
——. 2002. "Education for Sustainability: From Rio to Johannesburg; Lessons Learnt from a Decade of Commitment." Paris: UNESCO.
——. 2007. "The UN Decade of Education for Sustainable Development (DESD 2005–2014): The First Two Years." Paris.
——. n.d. "What Is Education for Sustainable Development?" http://en.unesco.org/themes/education-sustainable-development/what-is-esd.
UNFCCC. 2014. "Adptation." http://unfccc.int/focus/adaptation/items/6999.php.
UNICEF. 2012. *Climate Change Adaptation and Disaster Risk Reduction in the Education Sector: Resource Manual*, 217. New York: UNICEF.
——. 2015. "UNICEF Procedure for Ethical Standards in Research, Evaluation, Data Collection and Analysis." New York: UNICEF.
United Nations. 1992. "Agenda 21." Rio de Janeiro: United Nations Conference on Environment & Development.
——. 2015. "Transforming Our World: The 2030 Agenda for Sustainable Development." New York: United Nations.
United Nations Environment Programme. n.d. "Mitigation." https://www.unenvironment.org/explore-topics/climate-change/what-we-do/mitigation.
US Forest Service. n.d. "i-Tree." http://www.itreetools.org/. Accessed January 28, 2012.
Uzzell, D., E. Pol, and D. Badenas. 2002. "Place Identification, Social Cohesion, and Enviornmental Sustainability." *Environment and Behavior* 34(1): 26–53.

Valenzuela, A. 1999. *Subtractive Schooling: US-Mexican Youth and the Politics of Caring.* Albany: SUNY Press.

Van der Werff, E., and L. Steg. 2015. "One Model to Predict Them All: Predicting Energy Behaviours with the Norm Activation Model." *Energy Research & Social Science* 6: 8–14.

Van der Werff, E., L. Steg, and K. Keizer. 2013. "It Is a Moral Issue: The Relationship between Environmental Self-Identity, Obligation-Based Intrinsic Motivation and Pro-environmental Behaviour." *Global Environmental Change* 23(5): 1258–1265.

——. 2014. "I Am What I Am, by Looking Past the Present: The Influence of Biospheric Values and Past Behavior on Environmental Self-Identity." *Environment and Behavior* 46(5): 626–657.

Van Deth, J. W., B. Edwards, G. Badescu, A. Moldavanova, and M. Woolcock. 2016. "Associations and Social Capital." In *The Palgrave Handbook of Volunteering, Civic Participation, and Nonprofit Assocations*, edited by D. H. Smith, R. A. Stebbins, and J. Grotz, 178–197. New York: Palgrave Macmillan.

Van Zomeren, M., T. Saguy, and F. M. H. Schellhaas. 2013. "Believing in 'Making a Difference' to Collective Efforts: Participative Efficacy Beliefs as a Unique Predictor of Collective Action." *Group Processes & Intergroup Relations* 16(5): 618–634.

Vaske, J. J., and K. C. Kobrin. 2001. "Place Attachment and Environmentally Responsible Behavior." *Journal of Environmental Education* 32(4): 16–21.

Velasquez, A., and R. LaRose. 2015. "Youth Collective Activism through Social Media: The Role of Collective Efficacy." *New Media & Society* 17(6): 899–918.

Verplanken, B., and R. W. Holland. 2002. "Motivated Decision Making: Effects of Activation and Self-Centrality of Values on Choices and Behavior." *Journal of Personality and Social Psychology* 82(3): 434–447.

Veselý, M., P. Molenaar, M. Vos, R. Li, and W. Zeiler. 2017. "Personalized Heating—Comparison of Heaters and Control Modes." *Building and Environment* 112: 223–232.

Veselý, M., and W. Zeiler. 2014. "Personalized Conditioning and Its Impact on Thermal Comfort and Energy Performance—a Review." *Renewable and Sustainable Energy Reviews* 34: 401–408.

Videras, J., A. L. Owen, E. Conover, and S. Wu. 2012. "The Influence of Social Relationships on Pro-environment Behaviors." *Journal of Environmental Economics and Management* 63(1): 35–50.

Viscusi, W. K., J. Huber, and J. Bell. 2011. "Promoting Recycling: Private Values, Social Norms, and Economic Incentives." *American Economic Review* 101(3): 65–70.

Volk, T. L., and M. J. Cheak. 2003. "The Effects of an Environmental Education Program on Students, Parents, and Community." *Journal of Environmental Education* 34(4): 12–25.

Waite, S., A. Goodenough, V. Norris, and N. Puttick. 2016. "From Little Acorns . . . : Environmental Action as a Source of Well-Being for Schoolchildren." *Pastoral Care in Education* 34(1): 43–61.

Walker, B. H., and D. Salt. 2006. *Resilience Thinking: Sustaining Ecosystems and People in a Changing World.* Washington, DC: Island Press.

Wals, A. E. J. 2012. *Shaping the Education of Tomorrow: 2012 Full-Length Report on the UN Decade of Education for Sustainable Development.* Paris: UNESCO.

Wals, A. E. J., and A. Benavot. 2017. "Can We Meet the Sustainability Challenges? The Role of Education and Lifelong Learning." *European Journal of Education* 52(4): 404–413.

Wals, A. E. J., F. Geerlin-Eijiff, F. Hubeek, S. van der Kroon, and J. Vader. 2008. "All Mixed Up? Instrumental and Emancipatory Learning toward a More Sustainable World: Considerations for EE Policymakers." *Applied Environmental Education and Communication* 7: 55–65.

Wamuyu, P. K. 2018. "Leveraging Web 2.0 Technologies to Foster Collective Civic Environmental Initiatives among Low-Income Urban Communities." *Computers in Human Behavior* 85: 1–14.

Watts, R. J., and C. Flanagan. 2007. "Pushing the Envelope on Youth Civic Engagement: A Developmental and Liberation Psychology Perspective." *Journal of Community Psychology* 35(6): 779–792.

Wells, N. 2000. "At Home with Nature—Effects of Greenness on Children's Cognitive Functioning." *Environment and Behavior* 32: 775–795.

——. 2014. "The Role of Nature in Children's Resilience: Cognitive and Social Processes." In *Greening in the Red Zone*, edited by K. G. Tidball and M. E. Krasny, 95–110. New York: Springer.

Wells, N., and K. A. Rollings. 2012. "The Natural Environment in Residential Settings: Influences on Human Health and Function." In *The Oxford Handbook of Environmental and Conservation Psychology*, edited by S. Clayton, 509–523. Oxford: Oxford University Press.

WHO. 2018. "About WHO (World Health Organization)." http://www.who.int/about/mission/en/.

Wilkenfeld, B., J. Lauckhardt, and J. Torney-Purta. 2010. "The Relation between Developmental Theory and Measures of Civic Engagement in Research on Adolescents." In *Handbook of Research on Civic Engagement in Youth*, edited by L. R. Sherrod, J. Torney-Purta, and C. Flanagan, 193–219. Somerset, NJ: John Wiley & Sons.

Williams, C. C., and L. Chawla. 2016. "Environmental Identity Formation in Nonformal Environmental Education Programs." *Environmental Education Research* 22(7): 978–1001.

Wilson, E. O. 1984. *Biophilia*. Cambridge, MA: Harvard University Press.

Witt, R. S., E. S. Svendsen, and M. E. Krasny. 2018. "Civic Stewardship as a Catalyst for Social-Ecological Change in Detroit, Michigan." In Krasny, *Grassroots to Global*, 213–230.

Wollebæk, D., and P. Selle. 2007. "Origins of Social Capital: Socialization and Institutionalization Approaches Compared." *Journal of Civil Society* 3(1): 1–24.

Wolsko, C., and K. Lindberg. 2013. "Experiencing Connection with Nature: The Matrix of Psychological Well-Being, Mindfulness, and Outdoor Recreation." *Ecopsychology* 5(2): 80–91.

Wynes, S., and K. A. Nicholas. 2017. "The Climate Mitigation Gap: Education and Government Recommendations Miss the Most Effective Individual Actions." *Environmental Research Letters* 12(7): 074024.

Young, C. 2017. "The Impact of Narrative and Participatory Drama on Social Interactions, Attitudes, and Efficacy around Health and Environmental Issues in Malawi." PhD diss., Cornell University.

Yu, Y. 2018a. "Environmental Education in Campus Environmental Clubs in China." *Environmental Education* (Chinese) 6: 50–54.

——. 2018b. "Saving a Sense of Place, Saving Our Home." https://www.thenatureofcities.com/2018/06/03/saving-sense-place-saving-home/. Accessed October 15, 2018.

Zaff, J., M. Boyd, Y. Li, J. V. Lerner, and R. M. Lerner. 2010. "Active and Engaged Citizenship: Multi-group and Longitudinal Factorial Analysis of an Integrated Construct of Civic Engagement." *Journal of Youth and Adolescence* 39(7): 736–750.

Zeegers, Y., and I. F. Clark. 2014. "Students' Perceptions of Education for Sustainable Development." *International Journal of Sustainability in Higher Education* 15(2): 242–253.

Zelenski, J. M., and E. K. Nisbet. 2012. "Happiness and Feeling Connected: The Distinct Role of Nature Relatedness." *Environment and Behavior* 46(1): 3–23.

Zhang, H., E. Arens, M. Taub, D. Dickerhoff, F. Bauman, M. Fountain, W. Pasut, D. Fannon, Y. Zhai, and M. Pigman. 2015. "Using Footwarmers in Offices for Thermal Comfort and Energy Savings." *Energy and Buildings* 104: 233–243.

Zint, M. n.d. "My Environmental Education Resource Assistant (MEERA)." http://meera.snre.umich.edu/.

Permissions for Survey Instruments in the Appendix

Pro-environmental Behavior Survey, from Larson, Stedman, Cooper, and Decker 2015, 118, copyright © 2015 Elsevier Ltd., appears with the permission of Elsevier.

Environmental Action Scale, from Alisat and Riemer 2015, 19, copyright © 2015 Elsevier Ltd., appears by permission of Elsevier.

Environmental Knowledge Scales (system, action-related, and effectiveness knowledge), from Roczen, Kaiser, Bogner, and Wilson 2014, 972–992, appendix, copyright © 2014 Sage Publications, appears by permission of Sage Publications, Inc.

Systems Thinking Scale, from Davis and Stroink 2016, 577, copyright © 2015 John Wiley & Sons, Ltd., appears by permission of John Wiley & Sons, Ltd.

The Ultimate Cheatsheet for Critical Thinking, Global Digital Citizen Foundation, n.d., "Critical Thinking CheatSheet." from https://wabisabilearning.com/resources/?__hstc=163937376.0d760d2545ad4e63a96f4d5701e5ca02.1543426807224.1543426807224.1543482127182.2&__hssc=163937376.1.1543482127182&__hsfp=3919217888, is reprinted by permission of the Global Digital Citizen Foundation.

Values Scale, from Steg, Perlaviciute, van der Werff, and Lurvink 2014, 170, copyright © 2014 Sage Publications, appears by permission of Sage Publications, Inc.

New Ecological Paradigm, from Dunlap, Van Liere, Mertig, and Jones 2000, 433, copyright © 2000 The Society for the Psychological Study of Social Issues, appears by permission of John Wiley & Sons, Ltd.

Environmental Attitude Inventory, from Sutton and Gyuris 2015, 20–22, copyright © 2015 Emerald Group Publishing Ltd., appears by permission of Emerald Publishing Ltd.

Connectedness to Nature Scale, from Mayer and Frantz 2004, 513, copyright © 2005 Elsevier Ltd., appears by permission of Elsevier.

Connection to Nature Scale for Children, from Cheng and Monroe 2012, 41, copyright © Cheng and Monroe, appears by permission of Sage Publications, Inc.

Nature Relatedness Scale, from Nisbet and Zelenski 2013, 11, is published under Creative Commons agreement and appears courtesy of Frontiers in Psychology.

Sense of Place Scale, from A. Kudryavtsev, M. E. Krasny, and R. Stedman 2012, "The Impact of Environmental Education on Sense of Place among Urban Youth," *Ecosphere* 3, no. 4 (April): 29, http://dx.doi.org/10.1890/ES11-00318.1, an open access article published under the terms of the Creative Commons Attribution License, https://creativecommons.org/licenses/by/3.0/.

Political Efficacy (Internal and External), from Schulz and Sibberns 2004, 283, appears by permission of the International Association for the Evaluation of Educational Achievement.

Environmental Identity Scale, from S. Clayton and S. Opotow, eds., *Identity and the Natural Environment*, 61–62, copyright © 2003 Massachusetts Institute of Technology, appears by permission of The MIT Press.

Social Capital Survey for Youth, from Krasny, Kalbacker, Stedman, and Russ 2013, 1–23, © copyright 2013 Krasny, Kalbacker, Stedman, and Russ, was published by Routledge open access under the Creative Commons Attribution-Non-Commercial License, http://creativecommons.org/licenses/by-nc/3.0/.

Active and Engaged Citizenship, from Zaff, Boyd, Li, Lerner, and Lerner 2010, 743, copyright © 2010 Springer Nature, appears by permission of Springer Nature.

Wellbeing Scale, from McLellan and Steward 2015, 316, copyright © 2014 University of Cambridge, Faculty of Education, appears by permission of Taylor & Francis Ltd.

Climate Change Hope Scale, from Li and Monroe 2017a, 470, copyright © 2017 Li and Monroe, appears by permission of Sage Publications, Inc.

Climate Change Coping Scale, from Ojala 2012, 229, copyright © 2012 Elsevier Ltd., appears by permission of Elsevier.

Index

Page numbers followed by letters *f* and *t* refer to figures and tables, respectively.

Milton Keynes UK
Ingram Content Group UK Ltd.
UKHW021122290824
447565UK00005B/174

9 781501 747076